DR. CHAUNCEY STARR

PERGAMON INTERNATIONAL LIBRARY
of Science, Technology, Engineering and Social Studies

The 1000-volume original paperback library in aid of education,
industrial training and the enjoyment of leisure

Publisher: Robert Maxwell, M.C.

CURRENT ISSUES
IN ENERGY

THE PERGAMON TEXTBOOK
INSPECTION COPY SERVICE

An inspection copy of any book published in the Pergamon International Library will gladly be sent to academic staff without obligation for their consideration for course adoption or recommendation. Copies may be retained for a period of 60 days from receipt and returned if not suitable. When a particular title is adopted or recommended for adoption for class use and the recommendation results in a sale of 12 or more copies, the inspection copy may be retained with our compliments. The Publishers will be pleased to receive suggestions for revised editions and new titles to be published in this important International Library.

Other Pergamon Titles of Interest

CURRENT ISSUES IN ENERGY

A Selection of Papers

by

CHAUNCEY STARR

Electric Power Research Institute, California, USA

PERGAMON PRESS

OXFORD · NEW YORK · TORONTO · SYDNEY · PARIS · FRANKFURT

U.K.	Pergamon Press Ltd., Headington Hill Hall, Oxford OX3 0BW, England
U.S.A.	Pergamon Press Inc., Maxwell House, Fairview Park, Elmsford, New York 10523, U.S.A.
CANADA	Pergamon of Canada, Suite 104, 150 Consumers Road, Willowdale, Ontario M2J 1P9, Canada
AUSTRALIA	Pergamon Press (Aust.) Pty. Ltd., P.O. Box 544, Potts Point, N.S.W. 2011, Australia
FRANCE	Pergamon Press SARL, 24 rue des Ecoles, 75240 Paris, Cedex 05, France
FEDERAL REPUBLIC OF GERMANY	Pergamon Press GmbH, 6242 Kronberg-Taunus, Pferdstrasse 1, Federal Republic of Germany

First edition 1979

Reprinted 1980

British Library Cataloguing in Publication Data

Starr, Chauncey
Current Issues in Energy.
1. Power Resources—Social aspects—Collected works
I. Title
333.7   HD9502.A2   78-40695
ISBN 0-08-023243-4 (Hard cover) ✔
ISBN 0-08-023244-2 (Flexicover)

*Printed in Great Britain by
Biddles Ltd., Guildford, Surrey*

Contents

Foreword

During the years 1973 to 1978, when the ideas and views in this volume were being shaped and tested, Dr. Chauncey Starr served as the founding president of the Electric Power Research Institute, the research and development arm of the United States electric utility industry. As a new R & D organization, created to bring direction, balance and order to the sciences and technologies needed by the electric energy field, EPRI has become under Starr's aegis a focal point — the forefront of technical analysis and development — for the electric power industry.

During this same period, our whole society was becoming gradually aware that energy was no longer a mere parameter of our national health and welfare that could be treated in an *ad hoc* fashion; rather, it was the vital and central ingredient of our entire industrial structure, and of international development as well.

Accordingly, EPRI has set as its objective the planning and developing of comprehensive programs looking to the needs and technological options of the decades ahead, far into the next century. Chauncey Starr took a decisive and leading role in the structuring and direction of these programs; moreover, he exercised an intellectual leadership in sharpening perceptions of the essential issues facing us as we make decisions on the energy options that should be pursued with greatest vigor.

Each of us, in succession, while acting as Chairman of the Board of EPRI — William G. Meese, 1972 to 1973, James E. Watson, 1973 to 1975, Shearon Harris, 1975 to 1977, and Frank M. Warren, 1977 to present — has had the opportunity of working closely with Chauncey Starr. We have agreed to speak together in this foreword because we have been uniformly impressed with the depth of his analytical temper, with the force of his arguments, and with his articulation of the choices (and their potential consequences) that confront us in the electric energy field. We are glad, therefore, that these thoughtful and informative essays are now available to a wider concerned public.

The highlights of Dr. Chauncey Starr's career, his scientific, technical and management achievements, his honours and awards, are printed on page xi of this volume. They all testify to his strengths as a knowledgeable analyst of the energy concerns that confront us. We are joined in the hope that the ideas and positions developed by Chauncey Starr will prove useful in the setting of national objectives relating to energy and the quality of life — for us and future generations.

Chairmen of the Board,
Electric Power Research Institute
William G. Meese
James E. Watson
Shearon Harris
Frank M. Warren

Preface

The essays collected in this volume represent solid steps toward establishing a new "science of energy and society", a science about which, up till the present period, we have developed only the most rudimentary concepts. In fact, no such science structured from cause and effect relations yet exists, though some of its elements are beginning to emerge through the smokescreens of simplistic and idealistic notions that sometimes obscure our national dialogues.

These essays take a strong footing in the history of the role of energy in the evolution of society, and develop, through the modern tools of analysis and technological insight, views on the underlying problems regarding energy and its uses. In particular, they begin to articulate the connections between energy and production processes, between energy and economic output, and hence between energy and national and global welfare, both now and in the future.

Many serious questions, and many levels of questions, are raised and examined by the author, not in an academic or philosophical fashion, but to build effective guidelines for action. The author makes clear his conviction throughout that a range of technological options must be developed in the energy field if our future is to offer opportunities for continuing improvement of the human condition.

He argues, for instance, that energy may require a special analysis, and must not be treated from a classical economic viewpoint as just another economic variable in which demand and supply are fully responsive to price. Energy, he notes, may be different than most other goods, and more a basic necessity. Socially, therefore, we must do everything possible to insure an adequate future supply to prevent disastrous shocks and readjustments.

Some of the questions addressed in the papers:

- What is the true relation of energy consumption to Gross National Product?
- What are the energy options available to our society and what will be the consequences of the choices we make among them?
- Is it true that our industrialized society is using energy less efficiently than other societies?
- Will solar energy systems be able to compete effectively with other systems in the remainder of this century?
- What energy technologies are in sight, and what lead-times are likely in their development to commercial use?
- What evidence exists to define the relationship between energy use and the quality of life?
- How can social organizations evaluate intelligently the risks and benefits of different decisions when faced with a range of choices?

This listing of some of the issues examined by the author suggests his serious intentions. All in all, the exposition of these ideas forms a reasoned study of one of the most serious problems of our era or our stage of industrial progress. It represents, in fact, the view of a dedicated believer in technology and technological innovation as the answer to natural resource depletion; here in the late 1970s, we hear in Starr the voice of an unembarrassed believer in technology in the service of man and society. But in place of the naive belief that technology is the answer to all our problems, there is here a grasp on technology tempered by years of experience, which gives Starr's view a wholly different edge. Though he does not quite say it, Starr implies that, given the problems of our world — of its population growth, of its inexorable demand for resources — technology offers the principal means to their solution.

Technology may be, as Starr asserts, the *only* remaining unlimited resource available to man and to the kind of society he has evolved. It is this consciousness that gives these essays their grit, their determination, their unrelenting seriousness, and, ultimately, their readableness.

These essays were prepared with the invaluable assistance of members of the Electric Power Research Institute staff, including C. F. (Andy) Anderson, Dr. Walt Esselman, Dr. Chris Whipple, and Dr. Ed Zebroski.

About the Author

Dr. Chauncey Starr is vice chairman of the Electric Power Research Institute, a position to which he was appointed in May of 1978, following 5 years of service as the founding president of EPRI. From 1966 to 1973 he was the dean of the UCLA School of Engineering and Applied Science, following a 20-year industrial career during which he served as vice president of Rockwell International and as president of its Atomics International Division. In announcing Dr. Starr's academic appointment, UCLA Chancellor Franklin D. Murphy noted that "by training, by experience, and by performance, Dr. Starr is a unique blend of scholar-administrator. His interests encompass the full range from pure science to applied technology of the most sophisticated sort".

He received an electrical engineering degree in 1932 and a Ph.D. in physics in 1935, both from Rensselaer Polytechnic Institute in Troy, New York. He then became a research fellow in physics at Harvard University and worked with Nobelist P. W. Bridgman in the field of high pressures. From 1938 to 1941 Dr. Starr was a research associate at Massachusetts Institute of Technology in cryogenics.

He became associated with the Manhattan District in its early days at the Radiation Laboratory of the University of California at Berkeley, and subsequently at Oak Ridge. Following World War II, at Rockwell International he pioneered in the development of nuclear propulsion for rockets and ramjets, in miniaturizing nuclear reactors for space and in developing atomic electricity plants.

He is a member of the Energy Advisory Committee of the Office of Technology Assessment; a member of the US/USSR Joint Committee on Cooperation for Peaceful Uses of Atomic Energy; a member of the Energy Subcommittee of US – Israel Bi-National Advisory Council for Industrial Research & Development; a member of the US National Committee of the World Energy Conference; a foreign member of the Royal Swedish Academy of Engineering Sciences; and an "Officer" of the French Legion of Honor in recognition of his efforts aimed at promoting and furthering understanding between France and the United States in the field of scientific and industrial achievements.

He is most concerned with the two-way relationship between technology and society, and he has tried to clarify and strengthen this relationship on campus, in the community, in industry, and on the national scene.

Acknowledgements

Chapter 1

Fig. 1: C. Starr, R. Rudman, and C. Whipple, "Philosophical Basis for Risk Analysis". Reproduced, with permission, from the *Annual Review of Energy*, Vol. 1. © 1976 by Annual Reviews Inc.

Chapter 2

Text adapted from C. Starr, R. Rudman, and C. Whipple, "Philosophical Basis for Risk Analysis". Reproduced, with permission, from the *Annual Review of Energy*, Vol. 1. © 1976 by Annual Reviews Inc.

Fig. 2: C. Starr, "Social Benefit Versus Technological Risk", *Science*, Vol. 165 (Sept. 19, 1969), pp. 1232–1238, Fig. 4. Copyright 1969 by the American Association for the Advancement of Science.

Fig. 3: Adapted from "A New Concept in Risk Analysis for Nuclear Facilities", by P. McGrath, R. Papp, L. Maxim, and F. Cook, *Nuclear News*, Nov. 1974.

Fig. 6: US Atomic Energy Commission, *Reactor Safety Study: An Assessment of Accident Risks in US Commercial Nuclear Power Plants*, WASH-1400, 1974.

Fig. 9: E. J. Gumbel, *Statistics of Extremes* (New York: Columbia University Press, 1958), p. 237.

Chapter 3

Fig. 4: The Aerospace Corp., *Solar Thermal Conversion Mission Analysis*, Jan. 1975 (funded by National Science Foundation).

Figs. 9, 10: Honeywell, Inc., and Black & Veatch Consulting Engineers, *Dynamic Conversion of Solar-Generated Heat to Electricity*, Vol. II, Executive Summary, NASA CR-134723 (Washington, DC: National Aeronautics and Space Administration, Aug. 1974).

Fig. 11: Black & Veatch Consulting Engineers, *Solar Thermal Conversion to Electricity Utilizing a Central Receiver, Open Cycle Gas Turbine Design*, EPRI ER-387-SY (Palo Alto, Calif.: Electric Power Research Institute, March 1977).

Fig. 12: The Aerospace Corp., *Penetration Analysis and Margin Requirements Associated With Large-Scale Utilization of Solar Power Plants*, EPRI ER-198, Final Report (Palo Alto, Calif.: Electric Power Research Institute, Aug. 1976).

Chapter 4

Text reprinted from Vol. 39, *Proceedings of the American Power Conference*, 1977.

Chapter 5

Fig. 2: The Aerospace Corp., *Solar Thermal Conversion Mission Analysis*, Jan. 1975 (funded by National Science Foundation).

Chapter 6

Table 1, Fig. 1: United Nations, *Concise Report on the World Population Situation in 1970–75 and Its Long-Range Implications* (New York: U.N. Publications, 1974).
Fig. 2: From "The Human Population" by E. Deevey. Copyright © 1960 by Scientific American, Inc. All rights reserved.
Fig. 3: C. M. Cipolla, *The Economic History of World Population* (Pelican Books, 7th ed., 1978), p. 28. © Carlo M. Cippolla, 1962, 1964, 1965, 1967, 1970, 1974, 1978. Reprinted by permission of Penguin Books Ltd.
Figs. 4, 6–8: From "Energy and Power" by C. Starr. Copyright © 1971 by Scientific American, Inc. All rights reserved.
Fig. 5: From *Energy in the Future* by Palmer Putnam. © 1953 by Litton Educational Publishing, Inc. Reprinted by permission of Van Nostrand Reinhold Company.
Fig. 12: C. Starr and R. Rudman, "Parameters of Technological Growth", *Science*, Vol. 182 (Oct. 1973), pp. 358–364 Fig. 2. Copyright 1973 by the American Association for the Advancement of Science.
Figs. 13, 14: H. Darracott *et al.*, *Report on Technological Forecasting* (Washington, DC: Joint AMC/NMC/AFSC Commanders, 1967).
Fig. 15: M. S. Livingston and J. P. Blewett, *Particle Accelerators*. Copyright © 1962 by McGraw-Hill, Inc.
Fig. 16: G. Bernstein and M. Cetron, *Technological Forecasting*, Vol. 1, No. 38, 1969. Reprinted by permission of Elsevier North-Holland, Inc.
Fig. 17: Data from (*a*) Joint Committee on Atomic Energy, *Engineering and Scientific Manpower in the United States, Western Europe and Soviet Russia* (Washington, DC, 1956), and (*b*) S. Rhine and D. Creamer, *The Technical Manpower Shortage: How Acute?* (New York: National Industrial Conference Board, 1969).
Fig. 18: (*a*) Data for 1921–1960 are adapted from N. E. Terleckyj, "Research and Development: Its Growth and Composition", *Studies in Economics*, No. 28 (New York: National Industrial Conference Board, 1963); (*b*) data for 1965 are from US Bureau of the Census, *Statistical Abstract of the United States*, 1970.
Figs. 19–21: T. Gordon and A. Shef, *National Programs and the Progress of Technological Societies* (presented to the American Astronautical Society, Washington, DC, March 1968). Douglas paper 4964 (McDonnell Douglas Corp.).

Chapter 7

Fig. 1: From "Energy and Power" by C. Starr. Copyright © 1971 by Scientific American, Inc. All rights reserved.
Fig. 2: (*a*) Pre-1860 data are from Sterland, *Energy into Power* (Garden City: Natural History Press, 1967, pp. 120–121. © Aldus Books Ltd., London, 1967;

(*b*) 1860–1950 data are from P. Putnam, *Energy in the Future* (New York: Litton, 1953); (*c*) 1950–1965 data are from *Energy Supply and Demand Situation in North America to 1990* (Menlo Park, Calif.: SRI International, Nov. 1973).

Fig. 3: (*a*) Data for 1800–1950 are from P. Putnam, *Energy in the Future* (New York: Litton, 1953); (*b*) data for 1950–1975 are from US Bureau of the Census, *Statistical Abstract of the United States*, Sept. 1973.

Fig. 4: From *Energy in the Future* by Palmer Putnam. © 1953 by Litton Educational Publishing, Inc. Reprinted by permission of Van Nostrand Reinhold Company.

Fig. 6: R. Buckminster Fuller and John McHale, *World Resources Inventory: Human Trends and Needs*, Vol. 1, as reprinted in John McHale, *World Facts and Trends*, Macmillan, New York, 1972, p. 44.

Fig. 7: The Rand Corp., *The Impact of Electricity Price Increases on Income Groups: A Case Study of Los Angeles*, R-1102-NSF/CSA, March 1973.

Fig. 8: Reprinted from December 1974 issue of *Electrical World*. © Copyright 1974, McGraw-Hill, Inc. All rights reserved.

Fig. 9: J. Parent, "Some Comments on Energy Consumption and GNP", *Institute of Gas Technology Report* (Chicago: US Institute of Gas Technology, Oct. 1974).

Fig. 10: From "The Human Population" by E. Deevey. Copyright © 1960 by Scientific American, Inc. All rights reserved.

Fig. 11: From "Energy and Power" by C. Starr. Copyright © 1971 by Scientific American, Inc. All rights reserved.

Fig. 12: L. Mayer, "It's a Bear Market for Babies, Too", *Fortune*, Dec. 1974. Tom Cardamone for *Fortune* Magazine.

Chapter 8

Text reprinted from J. Dunkerley, ed., *International Consumption of Energy*, Research Paper R-10. © 1978 by Resources for the Future, Inc. A Resources for the Future Book published by The Johns Hopkins University Press.

Fig. 5: J. Darmstadter *et al.*, *How Industrial Societies Use Energy: A Comparative Analysis*, Preliminary Report. © 1977 by The Johns Hopkins University Press. A Resources for the Future Book published by The Johns Hopkins University Press.

Fig. 7: Copyright © 1975 by The Institute of Electrical and Electronics Engineers, Inc. Reprinted, by permission, from *IEEE Spectrum*, Vol. 12, No. 3, March 1975, p. 65.

Fig. 9: Reprinted from October 15, 1975, issue of *Electrical World*. © copyright 1975, McGraw-Hill, Inc. All rights reserved.

Chapter 9

Fig. 1: From "Energy and Power" by C. Starr. Copyright © 1971 by Scientific American, Inc. All rights reserved.

Fig. 5: National Aeronautics and Space Administration, Lewis Research Center, *Evaluation of Phase 2 Conceptual Designs and Implementation Assessment Resulting from the Energy Conversion Alternatives Study (ECAS),* NASA TM X-73515, April 1977.

Fig. 8: *Energy Technology to the Year 2000, A Special Symposium,* Oct./Nov. 1971. *Technology Review* (M.I.T.). © 1971 by the Alumni Association of the Massachusetts Institute of Technology.

Fig. 12: US Atomic Energy Commission, *Power Plant Capital Costs—Current Trends and Sensitivity to Economic Parameters,* WASH-1345, Oct. 1974.

Chapter 10

Figs. 1–3: "The Year 2000: Energy Enough?" *EPRI Journal,* June 1976, pp. 6, 8.

Fig. 8: 1947–1974 data from W. Dupree, Jr., and J. Corsentino, *US Energy Through the Year 2000,* rev. ed. (Washington, DC: US Bureau of Mines, Dec. 1975).

Chapter 11

Fig. 3: E. G. Cazalet *et al., Decision Analysis of California Electrical Capacity Expansion* (Menlo Park, Calif.: SRI International, Feb. 1977).

RISK-BENEFIT

1

*Risk-Benefit Analysis and Full Disclosure**

Introduction

All societies seek to maximize net social benefits through the allocation of available resources. The successful application of these resources to provide benefits depends on the quality of the decisions made. The purpose of risk-benefit analysis is to guide the decision-maker in these allocations. We face an almost continuous stream of such decisions, where the selection of a particular option in a decision can affect such things as our safety, our income, or the quality of our environment.

The key element in these decisions is uncertainty — it is always present. On one level are uncertainties with the available options; we cannot be certain that all reasonable options have been explored. On a second level is the uncertainty of the outcomes of the specific options under consideration. On a third level is uncertainty in the evaluation of net social benefit — we are not sure of the value system for judging benefits and social costs.

Despite these uncertainties decisions must be made; they cannot be delayed. To decide to postpone a decision is still a decision. To claim a lack of information begs the issue because information is always lacking to some degree. This applies also to the risk-benefit methodologies, flawed though they are; decisions are required and we must do the best we can.

The Decision Value System

The decision-maker varies from an individual to organizations representing major governmental functions. For individual choices, the value system used is that of the decision-maker; this greatly simplifies the decision process. In fact, we make intuitive risk-benefit analyses frequently; examples include the decision to replace tires or to buy a smoke detector. The impetus behind formal risk-benefit analysis is usually with decisions that are deemed "public issues". In these cases we face special problems not found in individual decisions or in all societies: decisions must satisfy a public consensus.

As examples of different risk value systems, two different approaches for defining the social cost of disability are illustrated in Fig. 1.[1] In this figure, the

---

*Presented at the American Association for the Advancement of Science, Washington, DC, February 13, 1978.

societal value system curve assumes a constant daily disability cost. This approach is taken by the US Public Health Service in decisions regarding their budget allocations. In this case the goal is to minimize disability days because the cost to society is assumed directly proportional to the total number of days of disability of the population.

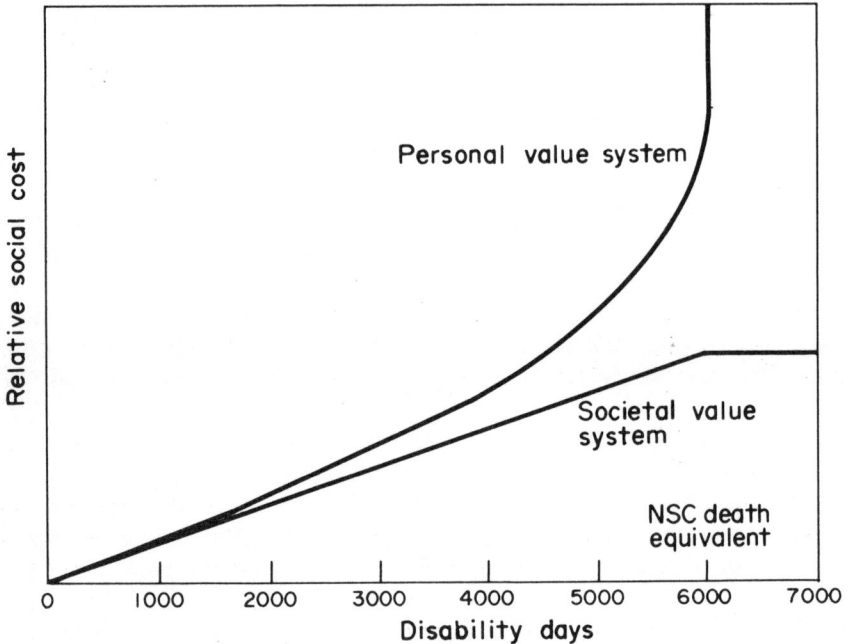

Fig. 1. Value systems for social costs of disability.

The second curve on Fig. 1 represents an individual's value system. Perhaps this explains the charitable organizations that raise money for research to develop cures for multiple sclerosis or muscular dystrophy. In terms of disability days to society, these diseases are probably far less significant than the common cold. The catastrophic nature of the effects upon the individual results in a personal evaluation of very high social cost.

One way to represent the difference between the two systems is to assume that personal value systems charge an increasing social cost to each successive disability day. As an example of this proposed representation of the personal value system, the social cost of one individual being disabled for a year is greater than the cost of 365 separate individuals each sick for one day. An example of such a representation would be:

$$\text{Total Social Cost} = NC_1 (1 + i)^t$$

where N is the number of individuals involved, i is a daily interest rate, t is the time in consecutive days of disability, and C_1 is the personal cost of one disability day.

The mix of personal and societal cost scales selected is dependent upon the decision-making agency. Autocratic countries and military organizations are more likely to use a societal scale because for these agencies the general welfare of society is placed above that of an individual. Democratic governments are more likely to use a mix of both scales; the congressional decision to spend extensively for cancer research is probably a result of the personal value systems of the mature voters.

In every social organization the decision-maker desires full knowledge of the risks and benefits embodied in the available choices. These risks and benefits, weighted by a system of values that defines the organization's tastes and willing-ness to make trade-offs, lead to a decision. A unique attribute of democratic organizations is that their value system cannot be institutionalized to the degree that it can in a highly disciplined organization. To illustrate the degree to which an institutionalized value system reduced the difficulty of decisions, note the speed and lack of debate that characterizes decisions by organizations with internal value systems (such as totalitarian governments, private corporations, and certain governmental groups such as those responsible for defense).

The purpose of making this distinction is not to praise dictatorial decision-making; it is a warning that we may back into an institutionalized value system as the path of least resistance. Although they are easiest to use, they are also likely to represent an organized minority that supports the institution rather than a public consensus. To the extent that risk-benefit methodologies can aid the decision process, the pressures to accept minority values are reduced.

The ability to make decisions rapidly has become an ideal in many organiza-tions. Strong pressures are felt by decision-makers to reduce the costs associated with delay, uncertainty, and rapidly fluctuating standards. In response to these pressures, assignment of decision, responsibility to institutions, and the adoption of an institutionalized value system may represent the path of least resistance to speedier decisions. But the social benefits gained by a reduction in delay comes at a terrible price. Even in the United States, institutionalized values are usually minority values; and the use of decision-making institutions is accompanied by a loss of public accountability. To the extent that full disclosure of risk-benefit analysis can reduce the time required for public consensus, the pressures to accept minority values are reduced.

Interpreting Majority Values

In a democracy the values implicit in national decisions should be the same values held by society in a majority sense. With this criterion we can examine the decision-making methods now used. First is the use of a direct popular vote. Although a public consensus is certainly achieved, this method is unwieldy for all but the most important issues. A complete list of the resource allocation decisions made annually would swamp the typical voter. Elections deciding risk-benefit issues are being used more frequently than in the past, as with the recent state-level votes on nuclear power, but a greatly expanded use of the popular vote would represent an enormous change from the *status quo;* I think it is prudent to look for stepwise improvements in the decision process. For example, I have neither the time nor appropriate technical background to sort through all aspects of all issues. I do not want to be the decision-maker, but I do want the decision-maker to consider my value system, as well as the values of others. Although this is a personal opinion, I think it reflects a majority view.

A second approach is inherent in our organization as a republic — let our elected representatives decide. While our institutions of elected officials are intended to be interpreters of values and opinions, they face the same problems as do individuals in considering these issues — there are too many issues to consider with the needed depth. Whether we are talking about Congress or local school boards, there are always decisions delegated to an administrative entity. This is the third approach. Only occasionally does the elected body specify the value system to be used in decisions — the Delaney Amendment (carcinogenic additives) is one example of such a value system. More often, the responsibility for interpreting social values is delegated to some government agency lock, stock, and barrel with the authority to make decisions. It is no surprise that bureaucrats find themselves in difficult positions — they must interpret social values with, at best, a vague mandate. This exposes administrators to second guessing by those who argue that a minority value system they do not like has been applied.

The incorporation of majority values into decisions would be relatively simple were those values known. But social values have proven extremely difficult to quantify, and their interpretation is highly subjective. As a first step, I believe a broad consensus exists in our culture for the following goals:

(1) meeting the basic material needs of all members of society;
(2) equality of opportunity in all areas;
(3) a clean environment;
(4) freedom from illness and risk; the right to a death from natural causes;
(5) basic personal freedoms (as in the Bill of Rights);
(6) availability of satisfying employment;
(7) security, both from wars and from crime;
(8) providing the needs of the elderly.

Undoubtedly this list is incomplete. Thus far, we have not achieved any of these goals in an absolute sense. Among these goals, allocations must be made that require difficult value judgements. We face choices between income and risk, between added police or more social security or reduced taxes; we choose between guns and butter. While the consensus goals are obvious, the trade-offs are not. It does little good for a regulator to know that we want safer cars and cheaper cars. That is not the choice; rather it is the balance between the two.

Although the objective of national decision-making is presumably to maximize the benefits to the public as a whole, the judicial system, in contrast, is more often concerned with the protection of individual and minority rights. Many public decisions that provide benefits on an aggregate level produce unfortunate distributional consequences. This equity issue is perhaps best described by the classic example: everyone in a city wants an airport, but no one wants it near his house. Often public decisions are halted or negated by the courts because of the conflict between societal benefits and individual rights.

This conflict between the national good and the rights of minorities is fundamental to human societies. The only societies in which this conflict has been resolved are totalitarian; the rights of individuals are always secondary to the interests of the state. In the US the balance between social goals and individual rights is continuously shifting through actions by legislative bodies and judicial decisions. This balance is itself representative of a public value system, one which must be incorporated into the decisions of public issues.

These considerations represent perhaps the most painful part of decision-making; and they certainly lead to the delays mentioned earlier. While I often wish that these issues could be resolved more rapidly, I recognize that not only does judicial review protect the rights of minorities, it provides a barrier against the institutionalization of minority values.

Methodology for Inferring Social Values

Thus far I have described the need for a risk-benefit methodology. We desire a method for interpreting consensus social values relating trade-offs between various social objectives. Risk-benefit analysis represents an attempt to provide a method for dealing with some of these trade-offs, those in which risks are significant factors.

Several years ago I described the uses of risk-benefit analysis as follows:[1]

"Three principal areas of application of risk-benefit analysis have been identified. The most fundamental application concerns the societal choice of expanding or curtailing the development of our existing technological capabilities. For example, the relative emphasis on mass and individual transportation systems is such an issue. A second area for risk-benefit analysis is setting performance targets for existing or new technologies where trade-offs between safety, environmental effects, and cost can produce major social and economic impacts — for example, air-quality criteria. The third application involves decisions between competitive technical systems that produce a similar beneficial function but with differing societal costs. For example, the selection of either coal or uranium as a fuel system for a new electrical power plant must be based on a comparison of their social costs, since the same output of electricity would be generated by both."

The underlying theme in these three different applications is an interpretation of the values relating to risk-taking. In 1969 I proposed a method[2] of interpreting these values through a revealed preferences methodology. The hypothesis underlying this method is that the social cost of risk is revealed through a measure of the benefit required for its compensation. Several interesting results came from that initial study: I found that risk acceptability did increase with benefits within certain ranges which might be considered as limits — risks below the natural-hazards level seem to be ignored, and risks above the average level from disease are rarely found for involuntary exposure. Another result of this study was the distinction between risks taken voluntarily and those over which the exposed have little control. My description at the time was,[2] "We are loath to let others do unto us what we happily do to ourselves". The extent to which this distinction represents a value system or is the result of perceiving less risk when we feel a degree of personal control is not clear.

This last comment points out a shortcoming in this method; past societal risk taking has been based upon perceived risks and benefits. We now have evidence that these perceptions do not correlate with statistical risk estimates.[3] Before we hastily reject this method on the grounds that it does not measure values but rather measures perceptions, we should recall that the same is true of all issues put to a vote. The success with which a society achieves its goals is determined in part by the degree to which perceptions reflect reality. To the extent that perceptions lag behind reality, resource misallocations will result.

Since that original paper was published, additional dimensions of risk have been identified as likely indicators of risk-taking values. One such risk dimension is the nature of the risk on a probability-magnitude space. It seems that chronic risks are preferable to catastrophic risks, even when the expected losses (probability times magnitude) are equal.

In one earlier paper, this point was discussed at length.[1] The explanation offered was an intuitive desire to maximize societal resilience, where resilience is defined as the ability of a society to recover from a damaging event with a minimum of social cost. In this study it was suggested that resilience could be modeled, and a form for such a model was suggested.

Our further work in this area has led to an improved understanding of the significant social measures relevant to resilience, subject to the qualification that any initial model should represent societal characteristics rather than the characteristics of the damaging event.

The first factor that contributes to resiliency is the population of a society; in effect the population sets a scale against which event magnitudes can be measured. If we wish to refine this term, the average lifetime of the inhabitants can be included as a product with the population; this would correct for the differences in expected loss between societies with different life expectancies.

Many high-consequence events pose risks over a specific area; a low population density reduces the likelihood that a sizeable percentage of a society will be affected by a single event. As a first cut, the area of a political subdivision can represent this factor; improvements to this measure would correct for highly nonuniform distributions of population within a region.

Both of these factors represent a static viewpoint of the capability of a society to absorb damage. Real societies respond to damaging events; the resilience to these events is directly dependent upon the repair processes that are initiated when damage occurs. The resources available for disaster response will obviously be some portion of the total societal wealth; as a first approximation the total wealth is probably a useful indicator of this ability.

A fourth social factor which influences resilience is the degree of technological development within a society. In our earlier paper[1] an increasing vulnerability was associated with the greater infrastructure of high technology societies; we now recognize a more complex relationship between technology and resilience. There is little doubt that societies with high levels of infrastructure depend upon more key links for vital services than do less technical societies. Associated with these key links are a number of failure modes; this gives rise to the apparent reduction in resilience. This is a probability related criterion — it is believed that extreme events are more likely in technical societies both as a result of technology itself and from the seemingly vulnerable infrastructure which accompanies it. While these negative attributes of technical societies have been discovered relatively recently, they have received considerable attention; in fact they may provide the basis for anti-technology philosophies. Less often considered are the beneficial characteristics of technology that apply specifically to resilience. It is clear that technical societies have a greater capability to respond to damage relative to nonindustrial societies. To explore this mechanism, we can consider cases in which vital portions of social infrastructure have been damaged. Examples include the bombing of German ball-bearing facilities and the loss of natural rubber to the US during World War II. Clearly both ball-bearings and rubber were vital to these wartime industrial societies, yet in both cases sufficient responses were made to compensate for the losses. In fact, we have learned to tolerate losses of major portions of infrastructure with minimal impact: labor strikes by railroad workers, dockworkers, autoworkers, or coal miners all have the potential to disrupt the flow of vital services, but we adjust to these occurrences as a matter of routine. The only way most of us were aware of the recent strikes which closed ports on both coasts of the US was because they were reported in the news. The impact upon the average consumer was negligible. It appears that the vulnerability associated with complex infrastructures is more imagined than real.

In addition to providing solutions to problems of its own creation, technology provides improved response to all types of damaging events. In many parts of the world, the typical aftermath of a flood is a cholera epidemic; this is not true of technical societies. It is precisely the same communications and transportation infrastructure that many see as evidence of vulnerability that in fact allows rapid response when a major accident occurs. In some cases, technology can reduce the damage from the event itself; we have hurricane-warning systems and earthquake building standards not found in less technical societies.

Our thinking in the area of resilience has in part been shaped by an analogy to the response of mechanical systems to perturbations. Through this analogy, we see the population of a society as representing an inertial term. The frictional forces in a mechanical system result in the dissipation of energy, and energy release is a characteristic of all damaging events. In social systems this dissipation is provided generally by natural actions that are area related — this certainly applies to the natural hazards (earthquakes, hurricanes) and the dilution of toxic materials (representing potential chemical energy) — and the use of area to represent this term seems quite appropriate. The restoring force in mechanical systems provides for the return to equilibrium, and the same is true for social organizations. An interesting result occurs if we wish to consider the integrated damage (that from event occurrence to repair) through an analogy to a simple linear second-order system with an impulse-driving function: in the overdamped mechanical system the time integral of displacement is equal to the event strength divided by the restoring force coefficient. If this is a valid analogy, it indicates that this integrated damage is greatly reduced as our ability to reduce event energies (such as through early detection and warning of risks) and our ability to respond to risks with effective repair mechanisms improves. If this is true, the technology is clearly a positive contributor to resilience.

Other research methods have been applied to the problem of interpreting a consensus risk-value system. The psychometric survey techniques (based on individual perceptions of risk) being used by Slovic, Fischhoff, and others have revealed additional risk dimensions[3] which influence risk perception, and are likely indicators of underlying risk values. Some of these factors are the degree of uncertainty associated with risk estimates, the newness of the risk, the likely severity of an event if one occurs (which is related to the frequency-magnitude topic described above) and the reaction to specific "dread" risks.

There are flaws with both the methods of revealed preference and psychometric survey; those of us using these methods are very aware of the shortcomings.[3] Without going into detail about these problems, I think it is fair to say that no one familiar with this field believes that a perfect method of inferring risk-taking values will ever be developed; rather we seek some broad descriptions of these values which would be applicable in the majority of issues. Even an imperfect analytical understanding of these values can be extremely useful in improving upon the intuitive interpretation that is now common practice.

The Decision Process

One should not confuse cost-benefit analysis with the decision process. The cost-benefit analysis in principle requires full disclosure of all the relevant aspects of any issue. Value system weighting of this analysis tends to give two separate options — one for the society as a unit, the other for individual members. The decision process incorporates all of this information but really focuses on a political optimization of the balance of individual, group, and societal interests.

My focus thus far has been with the interpretation of values necessary for allocations of societal resources — the other links in the decision chain can also be improved. First among these is the risk estimation. This analytical process has seen the most improvement of any of the steps in the process; in fact it is our improved ability to identify extremely low probability risks that is responsible for much current concern. Yet the procedures in practice are still often art rather than science: we can never be sure we have considered all significant factors, or correctly analyzed other factors (a quantitative description of the nonremovability of uncertainty is given in reference 1). Again, this begs the issue, because we face unquantified risks if we decide to postpone a choice.

The inference of social values has been discussed at length in the previous section. A comparison of the values implicit in many current decisions indicates[4] wide disparities in the intuitive interpretation — disparities that result in misallocations of resources.

The decision process itself is the only step for which a thorough methodology exists. While the existence of such a method (decision-making under uncertainty) does not guarantee its use, I suspect that those decision-makers who treat risks and values analytically will successfully combine the two factors. The successful incorporation of individual and social values into a single decision criterion represents the more difficult requirement for the decision-maker.

While these steps satisfy the formal requirements for a decision, I consider an additional step necessary in a democracy. The additional step is full disclosure of the decision-maker's interpretation of values and risks, including the risks of all options under consideration. Only through the feedback obtained from these disclosures do we avoid the institutionalization of minority values.

Full Disclosure

There is in the United States a tension between various organized minorities and the pluralistic interest as a whole. That various interest groups representative of minority value systems seek political influence is neither surprising nor troubling. What is troubling is the political strength gained by the claim that one group can represent majority values. In this respect I find the terms "public interest groups" and "consumer advocates" extremely misleading — we are all consumers and members of the public. The danger inherent in the actions of organizations representing special minorities is that a shadow government can exist. Full disclosure of all aspects of decision-making is a key element in assuring pluralistic choices.

A recent action serves as an excellent example. The Council on Environmental Quality (CEQ) recently recommended that plans for nuclear power development be delayed until uncertainties in radioactive waste management are resolved. To make such a recommendation responsibly, one should quantify the risk from proposed waste-handling procedures as well as is possible including a bound on the uncertainty bands surrounding the estimates, and to do the same for all other decision options. What was never mentioned by the CEQ was the risk associated with not having nuclear power — whether the risk is due to increased coal combustion, increased imports of energy, or risk from energy shortfalls. Had the CEQ, after analyzing all alternative risks, arrived at the same conclusion we would at least understand their decision process, if indeed there was one. Without these disclosures, public review of these decisions (something we all claim to want) is not possible. In this particular case I believe the recommendation, and the lack of substantive information that accompanied it, were direct results of the closed-door decision-making found when minority value systems are used.

Selective Thoroughness in Risk Analysis

The example just described serves to point up a shortcoming of virtually all decisions of this type — the risks associated with all decision options are not examined with equal thoroughness. When technological risk estimates are made all postulated risks are considered, even those at extremely low probability. Whenever possible, upper bounds on risk are determined, representing the uncertainty that is always present. The knowledge that significant risks could have been overlooked is always included in competent analyses. We do not treat the alternative choice (abandon the technology) in such a thorough manner. In general, the cost of giving up a technology is estimated on a "most likely" basis; low probability events are not considered.[5]

I can illustrate this with a simple calculation. Consider the nuclear-power examples described above; the costs of a nuclear moratorium are generally calculated by estimating the added cost for replacement fuels. One level removed are the second-order impacts usually ignored: uncertainty about energy availability leads to reduced capital investment, and in turn to fewer new jobs. Almost always ignored are the low probability non-nuclear accidents. Suppose the risk of a world war is one in a thousand per year (I do not know if that estimate would make me a pessimist or an optimist). If we remember the saber rattling about invading Saudi Arabia during the oil embargo, it seems likely that an energy shortage increases this probability, for example to 1 in 900. The result would be a 1 in 9000 increase in the chance of war (which must be multiplied by the probability of a shortage following a nuclear moratorium, viewed by many as close to unity). It is irresponsible that we expend money and manpower worrying about meteors striking power plants with infinitesimal probability, yet overlook this much more likely catastrophe. An added point in this example is that the second level impact — reduced jobs due to reduced investment — would never be publicly perceived as a consequence of the decision. So while there are social costs in our society that we are just now attributing to certain activities (such as the use of some chemicals) and there are unquestionably many other unidentified risks, we should recognize the potential for severe unattributable consequences through inaction.

Conclusion

The use of analytical risk estimates, of nonintuitive interpretations of social values relating to risk, and of full disclosure of the decision process, including risk estimates of all options, can lead to an allocation of resources preferable to the current practice, based on methods that have already been developed. The criterion I suggest for evaluation of risk-benefit analysis is the degree to which it represents an improvement over current practice, rather than its distance from a theoretical ideal. To this end, those of us in this field must be constructive in our criticism, always remembering that decisions are being made, resources allocated. Developments in this area will require many trial balloons; much research is needed to identify the areas that are ripest for improvement and to evaluate the performance of risk-benefit analysis versus intuitive decision making. Most importantly, a mandate for full disclosure must come from the very top of our government. We deserve to know in whose interest these decisions are being made, and we need assurances that all identifiable consequences of any action have been considered fully.

References

1. C. Starr, R. Rudman and C. Whipple, Philosophical basis for risk analysis, in *Annual Review of Energy*, Vol. 1, p. 629, 1976.
2. C. Starr, Social benefit versus technological risk, *Science*, **165**: 1232, 1969.
3. B. Fischhoff, P. Slovic, S. Lichtenstein and S. Read, How safe is safe enough? A psychometric study of attitudes towards technological risks and benefits, Draft, Oct. 1976, Decision Research, Eugene, Oregon.
4. R. Schwing, *Expenditure to Reduce Mortality Risk and Increase Longevity*, General Motors Research Laboratories Report GMR-2353, Feb. 1977.
5. E. Zebroski, Attainment of balance in risk-benefit perceptions, in *Risk-Benefit Methodology and Application: Some Papers Presented at the Engineering Foundation Workshop, September 22-26, 1975, Asilomar, CA*, UCLA report UCLA-ENG-7598, Dec. 1975.

2

*Risk and Risk Acceptance by Society**

Abstract

Various dimensions of risk are identified which relate to the manner in which risk is perceived and evaluated, and several self-consistent risk characteristics are explored. Factors which are thought to influence the perception of risk include the degree of personal control over the risk, the potential of episodic events, and the probable severity of injury if a risk event occurs. These factors contribute to a personal value system which differs from an expected value measure in that severe events are considered to be more costly than what an expected value measure would predict. Risk-benefit analysis can be applied to three problems: the allocation of resources for safety expenditures, the setting of standards, and societal risk-taking decisions. Calculations of benefit are needed for the third area of application; methods for the other two frequently do not require such a measure. The nature of the probability-magnitude relationship for a number of risks suggests that risk has certain repeatable characteristics. This relationship in particular is important in an estimate of risk impact on societal resilience, which is a measure of a society to withstand catastrophic events.

1. Introduction

All societies and individuals have recognized exposure to personal risk as a normal aspect of our life. Presumably such risk exposures have been accepted as necessary to attain a compensating benefit. A previous study[1,2] has explored this relationship and has suggested several hypotheses to explain the historically accepted balances of risk and benefits.

When individuals have "voluntarily" taken risks for personal pleasure or profit, they appear to be willing to accept relatively high risk levels in return for rather modest quantifiable benefits. For example, sportsmen frequently explore the outer physical limits of their chosen activity — with many resultant statistical accident records of their risk-taking propensities. Assuming that intangible benefits do exist, the controlling parameter appears to be the individual's percep-

---

*Presented at ENVITEC '77, Düsseldorf, Germany, 8-10 February 1977.

tion of his own ability to manage the risk-creating situation. If he believes he can do so, he is likely to take the chance.

The situation changes markedly when the individual no longer believes he can control his risk exposure. In such "involuntary" situations, the risk management is in the hands of a societal group usually remote operationally from the individual and his risk exposure. Major societal technical systems all create such "involuntary" risk exposures — as, for example, transportation systems, energy supply systems, public utilities, and food-supply systems. Whereas in the "voluntary" case the feedback loop of "control-risk-benefit-balance" is very tightly coupled by the individual, in the "involuntary" case this loop is usually very weakly coupled and dimly perceived, with each of its elements usually dispersed geographically, politically, managerially, and in time. Under these circumstances, the individual exposed to an "involuntary" risk is fearful of the consequences, makes risk aversion his goal, and therefore demands a level for such involuntary risk exposure as much as 1000 times less than would be acceptable on a voluntary basis. [1,2]

Inherent in all operational major technical systems is an implicit choice of such an acceptable level of risk for the involuntary exposure of the public. The above study has suggested the hypothesized relation shown in Fig. 1 between risk and benefit as a typical basis for such decisions. The figure shows the approximate relation between the per capita benefits to real income of a system and the acceptable risk as expressed in deaths per unit time (i.e. time of exposure in units of a year). The highest level of acceptable risks which may be regarded as a reference level is determined by the normal US death rate from disease (about 1 death/year per 100 people). The lowest level for reference is set by the risk of death from natural events — lightning, flood, earthquakes, insect and snake bites, etc. (about 1 death/year per 1,000,000 people).

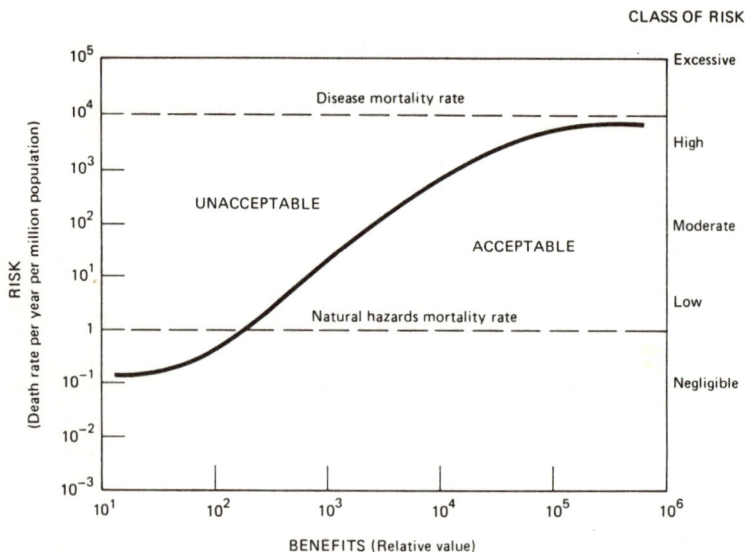

Fig. 1. Benefit-risk pattern, involuntary exposure.

Between these two bounds, the public is apparently willing to accept "involuntary" exposures — i.e. risks imposed by societal systems and not easily modified by the individual — in relation to the benefits derived from the operations of such societal systems.

In common familiar terms, an averaged involuntary risk may be considered "excessive" if it exceeds the incidence rate of disease, "high" if it approaches it, "moderate" if the risk is about 10–100 times less, "low" if it approaches the level of natural hazards, and "negligible" if it is below this. Events in these last two levels of risk have historically been treated as "acts of God" by the public generally — in recognition of their relatively minor impact on our societal welfare as compared to the effort required to avoid the risk.

Thus, any risk created by a new socio-technical system should be acceptably "safe enough" if the resultant risk level is below the curve of the figure in relation to its benefits. If, as is usually the case, a new technical system has a range of uncertainty in its risks, a design target may be set below the curve by an equivalent amount — possibly as much as a factor of 10 or 100.

Although the relationship hypothesized in Fig. 1 appears reasonable, its quantitative aspects should be considered as very tentative and primarily useful for illustrative purposes. The evaluation of the comparative benefits derived from the availability of technical systems is complex, difficult, and presently more heuristic than analytic. Because most new technical systems initially appear to be replacements for existing systems (for example, nuclear for coal power), public policy is generally concerned with comparative risk levels rather than comparative benefits. Perhaps it is not usually foreseen that new technology is likely to profoundly affect societal systems — the automobile became more than a substitute for the horse, although it was originally perceived only as a "horse-less carriage".

Thus, because comparative risk analysis is the issue of primary national concern, this paper will focus its discussion on the risk portion of the risk-benefit balance.

2. Evaluation of Risk

The study of risk analysis has as its objective the development of a methodology for predictive evaluation of future risks. Unfortunately, the literature on "futures" is apt to be obfuscated by a mix of personal value-system assessments of present trends and imaginative scenarios of alternative futures, usually written to dramatize the author's bias. For this reason, it is useful to recognize the existence of four different evaluations of future risk, as follows:

(1) *real risk*, as eventually will be determined by future circumstances when they fully develop;
(2) *statistical risk*, as determined by currently available data, typically as measured actuarially for insurance premium purposes;
(3) *predicted risk*, as analytically predicted from system models structured from historical studies;
(4) *perceived risk*, as intuitively seen by individuals.

Air transportation illustrates the above types of risks. To the flight insurance company, flying constitutes a statistically known risk; to the passenger purchasing the insurance at the airport, a perceived risk. In this case the perceived risk usually exceeds the statistical risk.* To the Federal Aviation Administration, proposed changes in air traffic patterns and equipment involves predicted risk as an approximation to future real risk.

2.1. Perception

In some cases, the risk which an assessor predicts through careful analysis of experimental and historical data, through the use of scientific laws, from transferring experience from related data, or a combination of these methods does not correlate well with either an individual or societal perception of risk. Perhaps the single most important factor in risk perception is risk manageability or controllability, where an individual feels safer if given some control over the degree of risk from an activity. Examples of this may be found in transportation where some individuals perceive more risk from flying than driving, or in skiing, when an occasional individual expresses more fear riding a chairlift up a mountain than skiing down. In both cases, the perception is not borne out by accident statistics. This is possibly the significant issue which results in voluntary activities having acceptable levels of risk several orders of magnitude higher than the levels associated with involuntary activities.[1] It is usually difficult to differentiate between a faulty perception of risk and its distortion by a personal value system which results in a conscious preference for a statistically higher risk, even when the risks are known as in sports.

An important factor in the perception of risk is the probable severity of injury if an accident were to occur. A complete assessment of risk requires that the potential effects of an accident be combined with a probability of occurrence. Because the accident probabilities are usually not given significant weight in an individual perception, concentration on dramatic accident consequences can lead to distortions of perception. Again the perceived safety of commercial air transportation suffers relative to the automobile (although the real risk is less) because the results of an average aviation accident are far more severe than the effects of the average automobile accident.

Another very significant factor is the episodic nature of some risks. On an individual level, a risk is usually perceived as a probability of death, with less severe effects such as disability, either temporary or permanent, considered insignificant in comparison to the risk of fatalities. On a societal level, the episodic preoccupation is evident in the common concern with disasters claiming many lives, while in fact single fatality accidents are responsible for the majority of all accidental deaths. It thus seems that the size of the potential accident is given more weight than the probability. This factor dominates the popular perception of risk from nuclear power plants, and can be recognized when the criticisms of nuclear power are examined, as the potential results of a major accident are frequently cited with no mention of the associated probabilities. This is probably

---

*The rate for life insurance at airports is $0.25 per $7500 insurance per flight. By 1971 data, this rate is approximately 30 times the actuarially fair value. Regular life insurance is usually available at somewhat less than twice the actuarially fair rate.

representative of societal values to a great degree, and activities capable of producing catastrophic accidents therefore must meet more stringent societal standards than higher frequency, individual risks. The failure to consider probability is, however, a perceptual factor.

One time-dependent effect that occurs in risk perception is the inconsistent discounting of future risk. The risks of smoking would seem far less acceptable if the risk were experienced immediately, rather than developed over a number of years. This perception does not seem to apply when nuclear power is discussed, probably because of the dramatized hypothetical scenario nature of the subject. Because the principal health effect of low-level radiation exposure is increased likelihood of cancer in the future, occurring perhaps 15 to 45 years after exposure, this risk should also be discounted. In most adversarial discussions of nuclear power, these are treated as immediate fatalities.

2.2. Perception of Benefit

Just as risk may be perceived to be different than the assessed value, so too may societal perceptions of benefit vary. In a previous paper,[1] an attempt was made to relate risk data to a crude measure of public awareness of the associated social benefits (see Fig. 2). The "benefit awareness" was arbitrarily defined as the product of the relative level of advertising, the square of the percentage of population involved in the activity, and the relative usefulness (or importance) of the activity to the individual. These assumptions may be too crude, but Fig. 2 does support the reasonable position that advertising the benefits of an activity increases the public acceptance of a greater level of risk. Smoking has been cited as an example of a perceived benefit due in large part to advertising.[3] The decision to ban cigarette advertising on television indicates a tacit understanding of this phenomenon.

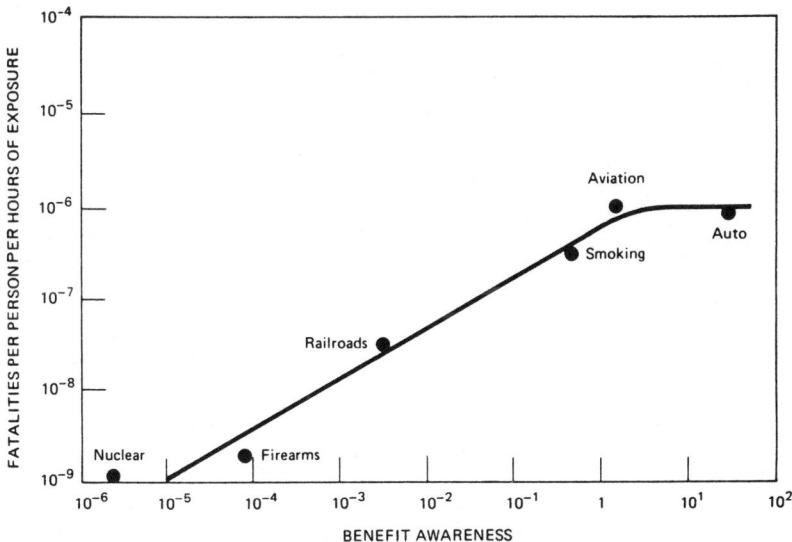

Fig. 2. Accepted risk versus benefit awareness.

2.3. Perception versus Reality

When a gap appears between the perceived and analytically assessed risks and benefits, the societal decision-makers are called upon to resolve the conflict. A recent example of such a resolution was the requirement that automobiles be equipped with a seat-belt interlock system. In this case, the individual perception underestimated the risk, as the majority of people were not using seat belts. The US regulatory decision was made on the basis that the reduction in risk was substantial enough to justify the cost of the safety system. This regulation was subsequently repealed, as it seemed to the public that the total social cost of the system was underestimated: the cost of inconvenience was apparently significantly large, thus the repeal.

No general method of dealing with differences between perception and technical assessment exists in the societal decision-making process. This problem is not directly within the realm of the technologist, as the technologist is not, in general, the decision-maker. The role of the technologist is to provide an assessment of risk and benefits. The societal perceptions and values are made visible through normal political inputs. The technologist should, however, indicate to the decision-maker that reliance on societal perceptions will usually result in a misallocation of resources. The difficult task of separating faulty perceptions from societal values has not been solved.

Situations frequently arise where the individual member of society simply does not have the information to relate perception to reality. As a case in point, we may consider the perception of the solar system. To those who are not astronomers, the visible evidence is that the sun travels around the earth. The only reason that the educated population as a whole believes otherwise is because the word of the scientists was accepted. What seems significant about this situation is the duration of the perceptual lag; that is, the time between analytic discovery and cultural acceptance. While this example may seem amusing in its historical context, the same sort of situation exists today in other areas.

Three levels of perceptual problems can be identified. First, there are the real-time problems of perception. These include the factors previously identified as controllability, episodic preoccupation, and other similar effects. The second is the perceptual lag problem. In general, perception eventually catches up with reality, but the time required may result in a massive misallocation of resources in the interim. The third level is assessment of hypothetical future risks. If decisions were based entirely upon public perception, many technologies could not have been introduced because an exaggerated fear of new risks would have prohibited their introduction.

2.4. Societal Value System for Risk

The incorporation of risk estimates into national decision-making is in its early stages of methodology development. The use of a societal value system for risk acceptance must rank as the major unresolved issue in this part of the decision process. While methods have been proposed,[4] none has yet been put into practice. Figure 3, taken from this reference, is representative of a proposed method of dealing with this issue.

The particular method referred to in Fig. 3 is called utility theory, and briefly stated, its estimates measure the social values associated with various objectives on a common scale. In theory, the diverse social costs associated with energy production, and with a lack of energy, may be estimated for comparison. The key question in such a method is that of estimating a societal utility function. This process occurs in a nonempirical manner at the present time. The social value system is made known through various channels of public expression such as elections, letters to congressmen, and the many polls. Such a process is incapable of separating values from perceptions.

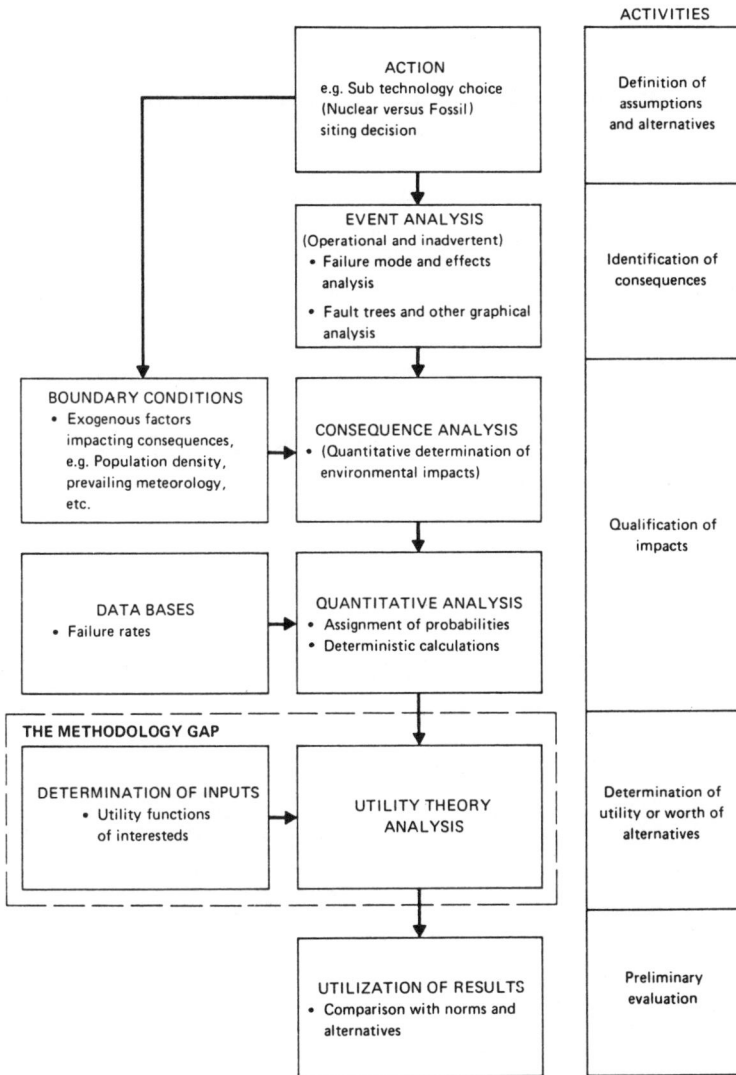

Fig. 3. Illustrative flow chart: risk analysis.

2.5. *Use of Risk Data*

A decision-maker is frequently called upon to compare different types of risk (and more generally, different types of social cost) resulting from the various options under consideration.

At one end of the spectrum is the social cost function that follows expected societal value rather closely. This approach is taken by the US Public Health Service in decisions regarding their budget allocations. In this case, the goal is to minimize disability days, because the social cost is assumed directly proportional to the total number of days of disability.

A second type of social cost function results if an individual's value system is used. This perhaps explains the charitable organizations that raise money to research cures for multiple sclerosis or muscular dystrophy. In terms of disability days to society, these diseases are probably far less significant than the common cold. The catastrophic nature of the effects upon the individual results in a personal evaluation of a very high social cost.

One way to represent the difference between the two systems is to assure that personal value systems charge an increasing social cost to each successive disability day. As an example of this proposed representation of the personal value system, the social cost of one individual being disabled for a year is greater than the cost of 365 separate individuals each sick for one day. The representation would be:

$$\text{Total Social Cost} = NC_1 (1 + i)^t \tag{1}$$

where N is the number of individuals involved, i is a daily interest rate, t is the time in consecutive days of disability and C_1 is the personal cost of one disability day. If such outcome is considered rather than a probability of outcome Fig. 4 represents such a curve.

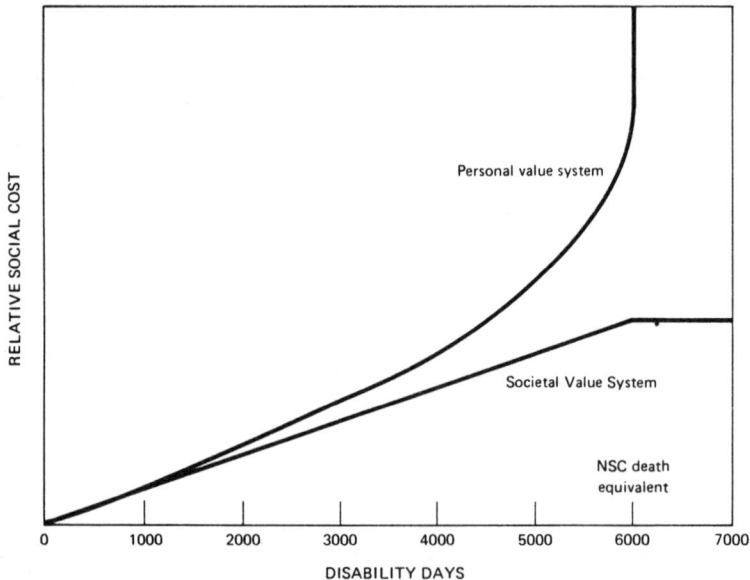

Fig. 4. Value systems for social costs of disability.

The mix of personal and societal cost scales selected is dependent upon the decision-making agency. Autocratic countries and military organizations are more likely to use a societal scale, for the general welfare of society is placed above that of an individual. Democratic governments are more likely to use a mix of both scales; the US congressional decision to spend extensively for cancer research is probably a result of a personal value system.

The choice of a cost system will, in some cases, determine whether an option with a high probability of a few fatalities is preferable to an option with a lower probability of a greater number of fatalities. While there are compelling arguments for both systems, the choice can only be made by a reading of the societal preference.

The term "value of life" appears frequently in the risk-benefit literature, and the meaning is derived in a variety of ways.[5] The original approach was to discount future earnings at an appropriate rate, to arrive at some value. Such a method has obvious drawbacks, for the life value of retired persons goes quickly to zero as does the value for those who work without direct compensation (e.g. housewives). Court awards for accidental deaths are not representative, for the compensation to an individual's survivors is probably representative of lost earnings. The approach taken by Sagan[6] is to charge a flat rate ($50) per day of disability, and to use the National Safety Council equivalent of 6000 disability days for a death, which gives a life value of $300,000. This is a valid method if the societal value system is the one used but would not work for the personal system.

2.6. Comparison of Risk and Benefit

It is rare to find a technology that provides a "new" benefit. The high technology energy and transportation systems frequently used as examples in risk-benefit studies always are competitive with other energy and transportation technologies. When a "new" technology arises, for example a medical treatment for a previously incurable condition, the risk-benefit analysis weighs the social cost against the benefits to provide insight into the criteria for acceptability.

A more common procedure is to compare several technologies through a risk-benefit viewpoint. If the risks associated with nuclear power are analyzed, it is more common to weigh these risks against the social costs of using coal or oil than to balance risks and benefits. The reasons for this is simple. Coal and oil have been acceptable fuels for power plants for years. If the social costs of nuclear power compare favorably with those of coal and oil, the nuclear power automatically satisfies the risk-benefit criteria. Under these circumstances, the concept of benefit remains simple: the benefit is expressed in terms of electrical energy. Such a situation proves enormously helpful in an assessment of an energy technology, because no widely accepted method of assessing benefits exists. The typical problems of assessing benefits of energy, and electrical energy in particular, stem from the fact that prices are set according to costs, and a free market condition does not apply.[7] Attempts to improve the benefit assessment methodology frequently include additional benefits such as tax revenues and local employment, but these techniques usually represent double accounting. An alternative method from classical welfare economics is to use the concept of the consumer surplus, but such a method produces severe measurement difficulties. Fortunately, an assessment of benefits is not necessary for the general analysis of energy technologies.

A commonly employed method in comparison of energy technology is predicting the health costs from various types of power plants with the assumption that each plant is of a given power rating and the plants are sited at equivalent locations. In this case the benefits are truly identical: the generation of power to serve the needs of a given population. The quantification of social costs may be simplified under these circumstances, because all social costs need not be translated to a dollar amount. Instead, each technology is responsible for various types of health effects which may be expressed in more convenient units such as injuries, fatalities, disability days, or amounts of life shortening. Due to the societal distinction between voluntary and involuntary risk, the analyses are frequently divided into occupational and public health effects. A risk-benefit analysis includes all the social costs associated with a given technology.

The methodology of assessing social costs due to environmental burdens is less developed than the comparative health cost methods. The reason for this is that the environmental burdens are of an aesthetic nature, while health costs may usually be translated into medical costs, days of disability, and fatalities. For this reason, risk-benefit studies have focused upon the health effects associated with the production of power, with less detailed comments offered upon the environment burden.

An even more difficult externality to quantify in the energy field is the potential social cost of undependable supplies of a particular form of energy. The current emphasis applied to reducing oil imports, and establishing a goal of energy self-sufficiency, indicates that this externality is now getting recognition. On a smaller scale, some industries are now switching from natural gas to electricity, due to the threat of nonavailability.

2.7. Use of Risk-Benefit

Risk-benefit analysis can be applied in three principal areas. The first area relates to the allocation of resources for safety expenditures. In general, a law of diminishing returns applies to all types of controllable risk — risk reduction becomes progressively more expensive. If an accepted methodology for determining the social cost of risk is developed, safety allocations would be determined as demonstrated in Fig. 5.

Ideally, safety expenditures should be made until the marginal exchange between social cost and control cost are equal, that is, the expenditure of one dollar for safety is expected to reduce the social cost by one dollar. At this point, the total cost is minimized.

The second area of application for risk-benefit analysis is in the setting of standards, whether for occupational safety or for public health risks. This will provide an analytic basis for trade-offs such as that between automotive emissions and mileage, and electric power plant emissions and electricity cost.

The final and most important application is in societal risk-taking decisions. Risk-benefit analysis provides a background of consistency through which decisions can be carefully examined. The consistency produced will ensure that the path of minimum social cost will be selected.

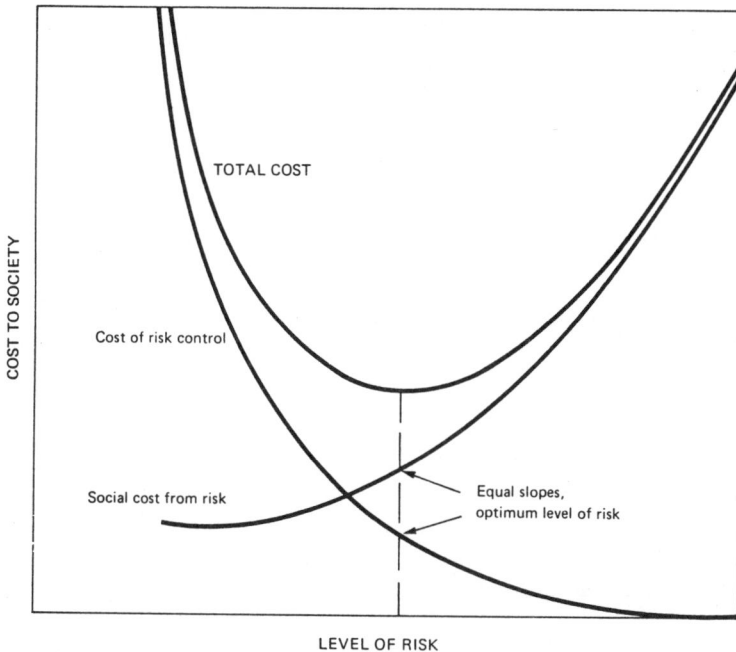

Fig. 5. Social costs and control costs versus level of risk.

3. Nature of Risk

Before a methodology for dealing with risk on a societal level can be developed, it is first necessary to examine the nature of risk and the history of societal risk taking. The preceding outline covers the key points in such a process, and each major element will be described.

Risk is here defined as the probability per unit time of a unit cost burden occurring. The cost burden may be measured in terms of injuries (fatalities or days of disability) or other damage penalties (expenses incurred) or total social costs (including environmental intangibles). Risk thus involves the integrated combination of (1) the probability of occurrences, (2) the spectrum of event magnitudes, and (3) the spectrum of resultant personal injuries and related costs.

As thus defined, risk is directly related to the familiar insurance premium determination for any class of activity. Thus, automobile insurance premiums are based on the frequency of collisions, the distribution of their physical magnitude, and the distribution of the resultant damage. Better roads were addressed to the first point, elastic bumpers to the second, and seat belts to the third. This separation of parameters is often glossed over in many discussions of the subject, but it is important for analytic studies.

3.1. Characterization of Risk – Exposure Probabilities

The probability of exposure to risk (the first item above) has three major elements: space, population, and time dependency.

The spatial character of exposure probabilities can range from quite localized risks to global hazards. For example, the production and utilization of energy causes effects over a wide range of space; hydroelectric dams and some pollutant emissions represent localized risks, whereas the production of carbon dioxide, the increase in global turbidity, and nuclear safeguards are of global concern.

A second aspect of exposure probability is determined by population dependency. For many types of risk, specific groups may be identified as the population fraction that bears a specific risk. While many such groups exist, certain group characteristics may be identified as the determinants of the risk-bearing population. The following list includes some types of group dependent risks:

(1) genetic diseases;

(2) age — childhood or geriatric diseases. Certain accident probabilities are higher for certain age groups, such as poisons (children), falls (older persons);

(3) occupational — pneumoconiosis (black lung disease) among coal miners, burns to fire fighters, and many other occupational hazards;

(4) sex — incidence of heart attacks in males, risk to women in childbirth;

(5) activity groups — smokers, skiers, and gun owners.

The time-dependent exposure probability may be classed into three major groups:

(1) continuous — describes most diseases and accidents which occur any time;

(2) periodic — certain risks are periodic in nature such as hurricanes, tornadoes, influenza, and some earthquakes;

(3) cumulative — the exposure probability of risk is in some cases related to previous exposure such as smoking, pneumoconiosis, asbestosis, and radiation effects.

3.2. Event Magnitude and Impact

The magnitude of a risk event is a physical measure of the event, and does not consider the consequences of an event to individuals. In this sense, event magnitude is not directly a measure of risk, but rather of event characteristics. Examples of event magnitude measures are earthquake magnitude, flood height, or explosion energy release.

Event impacts measure the effects of risk events upon exposed populations. As such, the impact is a key part of risk (the other is probability). This distinction between magnitude and impact is made because risk estimates frequently come from a combination of two estimates. The first is the probability of an event of a given magnitude, the second is the impact of an event versus magnitude. From these two estimates, the risk (probability of a specified impact) may be estimated. The relationship between magnitude and impact is explored in the next section.

4. Hypothesized Laws of Risk

Risk has certain usually repeatable characteristics which may be termed laws of risk. An exploration of these characteristics is useful for the purpose of comparing risks from different sources.

4.1. Logarithmic Relationship between Probability and Magnitude

Risk is properly defined as the probability per unit time of a unit cost burden (injury) occurring. It consists of both the probability that some event will occur and the magnitude of damage that results if the event occurs. From data for various types of risk (see Figs. 6 and 7) it is clear that the probability and

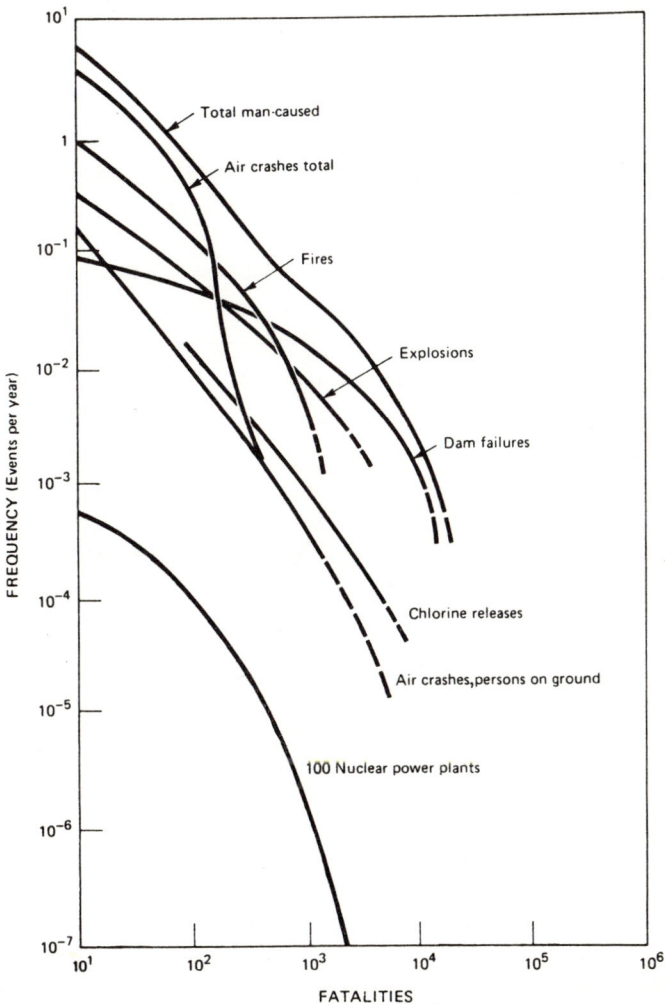

Fig. 6. Frequency of fatalities due to man-caused events.

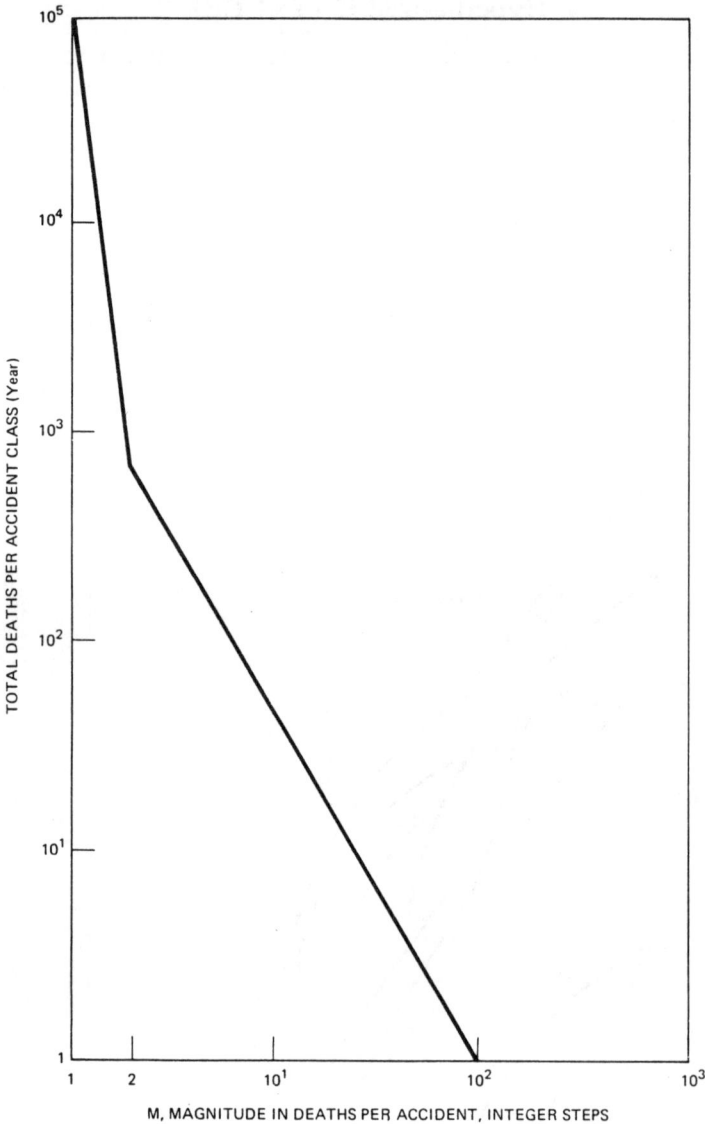

Fig. 7. Total deaths arising from accidents which kill M people versus magnitude.

magnitude are not independent but that for many risks the high impact events are much less common than the lower impact events. This result is in itself not surprising where societal choice existed, for it seems clear that in response to both natural and man-made hazards, successful societies have selectively evolved in a manner designed to avoid catastrophic events. What is surprising is that even for natural phenomena (earthquakes) usually the probability (or frequency) of the event falls off at least as fast as an exponential function of the magnitude for many types of risks (see Figs. 8 and 9). Mathematically this is:

$$\log f = a - bm \qquad (2)$$

where f is the frequency, m the magnitude, and a and b are constants.

If estimates of the probability of high magnitude events are desired, the data base is usually sparse, because the events are rare. The probability distribution for the general type of risk can, however, be estimated from the available data for

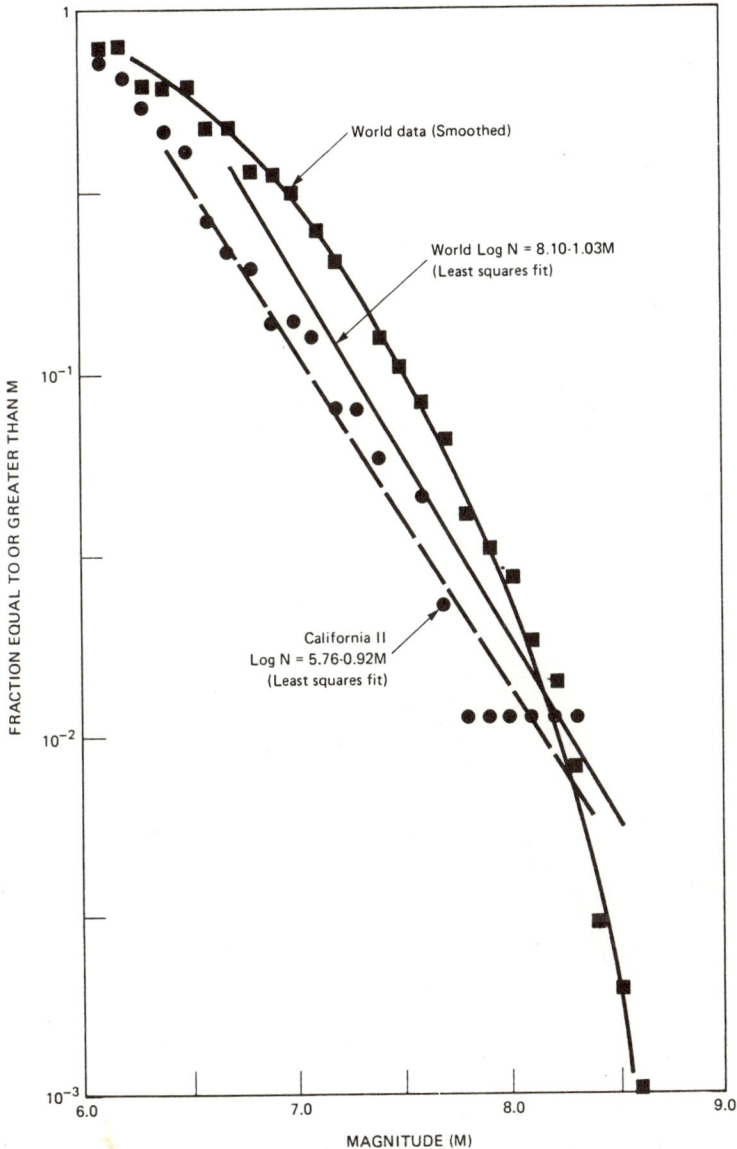

World data (Smoothed)

World Log N = 8.10-1.03M
(Least squares fit)

California II
Log N = 5.76-0.92M
(Least squares fit)

FRACTION EQUAL TO OR GREATER THAN M

MAGNITUDE (M)

Fig. 8. Magnitude distribution of shallow earthquakes, 1904-1952.

CURRENT ISSUES IN ENERGY

less extreme cases. A class of statistical distributions, known as asymptotic distributions, have proven useful in estimating extreme probabilities for natural hazards such as floods and earthquakes. The asymptotic distributions come from the theory of extreme value distributions,[8] and are useful because less restrictive conditions are required for the use of the asymptotic distributions than for the exact distributions of the extremes of functions.[9] Extreme value distributions deal with the distribution of the maximum or minimum values in large samples of

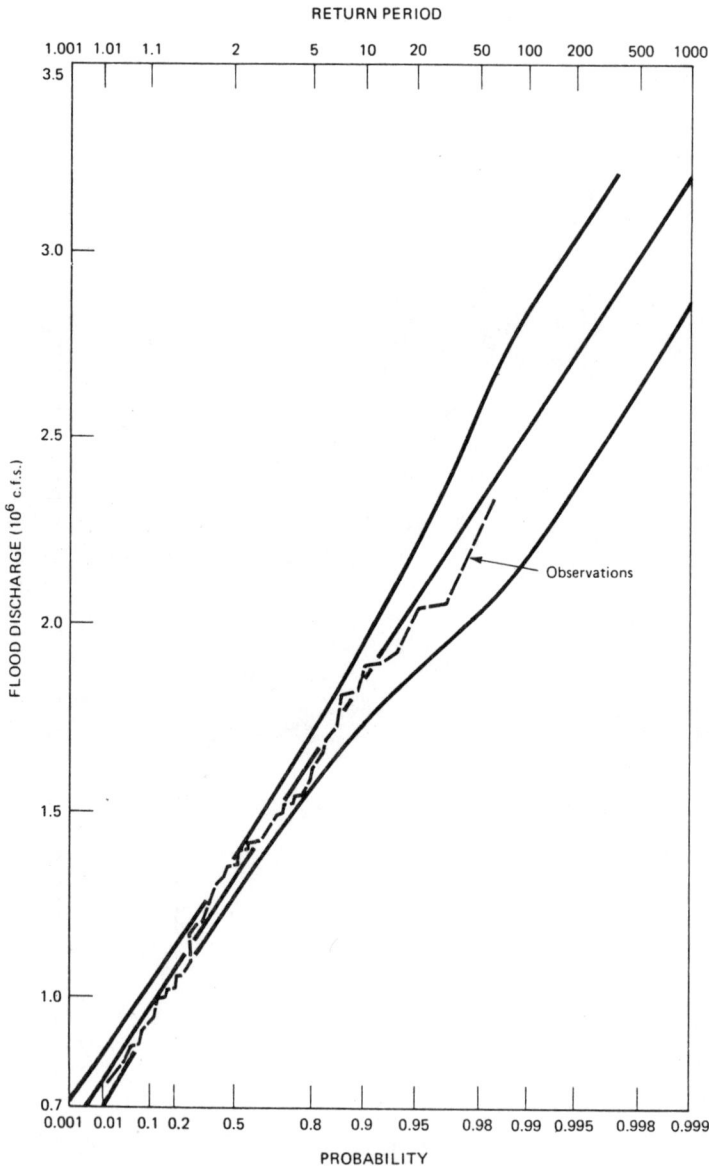

Fig. 9. Mississippi River floods, Vicksburg, Mississippi, 1898-1949.

independent values drawn from an initial distribution. This approach requires knowledge of the initial distribution and the sample size. This type of distribution is frequently applied to floods;[8, 9] the nature of the result may be seen from Fig. 9.

4.2. Resilience

The concept of resilience to a particular event impact was introduced briefly in earlier sections, in particular in the discussion of the elasticity of individuals to the hazards of drowning or electric current. This concept can be expanded to the societal level to consider those events capable of causing a lasting social impact.

Two principal types of societal resilience should be examined, relative to two different types of damage. First is the accident such as an earthquake or flood which causes a great deal of immediate damage. The threat associated with this type of event is that the loss of substantial portions of a population and resources will require a substantial effort and time for recovery. The second type of risk to a society is that which threatens the societal structure and culture, rather than large losses of life. This second type of risk and its relations to societal resilience is of great importance, for large-scale physical disasters have been routinely tolerated by societies for years, and populations continue to live in areas of high risk from flooding or earthquakes.

The importance of the resilience factor is the apparent nonlinearity of societal response to accidents as the impacts increase. The following quotation perhaps dramatizes the issue:

> "Small accidents throughout the world kill about 2 million people each year, or 4 billion people in 2000 years. This is "acceptable" in the sense that society will continue to exist, since births continually replace the deaths. But if a single accident were to kill 4 billion people, that is, the population of the whole world, society could not recover. This would be unacceptable even if it only happened once in 2000 years."
>
> (quoted from Wilson and Jones[10])

The concept of resilience deals primarily with the survivors' point of view, unlike individual risk acceptance.

This study of societal resilience is an approach to evaluation of the elasticity of a society to high impact events and to the identification of those characteristics of societies which determine such resilience. One such factor is the level of technological development of the society. The more technically advanced a society is, the lower its resiliency. This is due to a complex societal infrastructure which cannot operate with a sizable portion of its subsystems missing. In less developed areas, with less divisions of labor, the culture is more immune to accidents. An example of the lesser resilience of technological cultures may be seen from the bombing of ball-bearing factories in Germany in World War II. While the ball-bearing industry was small in comparison to the total industrial capability, the infrastructure was such that the loss of ball bearings would have been a substantial blow to the society at that time. Conversely, the severe flooding that caused over 100,000 fatalities in Bangladesh was not a major test of societal resilience, because the social structure is dependent only upon the quantity of food produced, not upon vital industries or key population segments. In such a case, the population rapidly recovers to a level constrained only by the food supply. A striking example of the sensitivity of the US economy occurred with the oil embargo of 1973-74, and the effects were far reaching even though the imported oil represented only a small portion of the total economic expenditure of the United States.

The key factors which seem to determine the societal resilience to a given accident or event may be arranged in the following dimensional form; if the resilience is denoted as R, then:

$$R = \frac{PL}{B\,(T)\,S^a t_r^{\,b} t_d^{\,e} M^f t_c^{\,g}} \tag{3}$$

seems reasonable. In this relation, P is the societal population, L its average lifetime, B a monotonically increasing function of the level of technology T, S the size of population affected by the event, t_r the average recovery time for an injured population group, t_d the time to discover the source of damage, t_c is the time to correct the situation, and M the average disability per person affected. The factors a, b, e, f, and g are constants. While there are unquestionable other factors which are significant, this relationship may provide a starting point for work on the subject of societal resilience. Ecologists have studied a similar problem, that of species resilience.[11]

From a societal welfare view, as contrasted with that of an individual, it appears that the "societal resilience" concept may be the most significant overall measure of acceptable risk associated with technical systems. Much more study will have to be done before such a measure can be usefully applied. However, even at this early stage it is clear that the prime criterion in any decision will always be the dominant national attitude in the philosophic dichotomy arising from the welfare of the state versus the welfare of the individual.

References

1. C. Starr, Social benefit versus technological risk, *Science*, **165**: 1232 (19 Sept. 1969).
2. C. Starr, Benefit-risk decision making, *Nat. Acad. Engng*, April 1971.
3. J. S. Tamerin and H. L. Resnik, Risk taking by individual option — case study: cigarette smoking, *Nat. Acad. Engng*, April 1971.
4. P. E. McGrath, R. Papp, L. D. Maxim and F. X. Cook, Jr., A new concept in risk analysis of nuclear facilities, *Nuclear News*, Nov. 1974.
5. R. Zeckhauser, Procedures for valuing lives, Discussion Paper Series, Public Policy Program, Kennedy School of Government, Harvard University, Number 29D, Second Version, June 1975.
6. L. A. Sagan, Human costs of nuclear power, *Science*, **177**: 487 (11 August 1972).
7. M. Marcus, The economic benefits of nuclear power plants, *Public Utilities Fortnightly*, (20 June 1974) p. 26.
8. E. J. Gumbel, *Statistics of Extremes*, Columbia University Press, 1958.
9. G. E. Apostolakis, *Mathematical Methods of Probabilistic Safety Analysis*, UCLA report UCLA-ENG-7464, Sept. 1974.
10. R. Wilson and W. J. Jones, *Energy, Ecology, and the Environment*, Academic Press, 1974.
11. C. S. Holling, Resilience and stability of ecological systems, *Ann. Rev. Ecol. System.*, **4**, 1973.

ENERGY TECHNOLOGY

3

Role of Solar Energy in Electric Power Generation*

Introduction

Solar energy can affect future electric generating systems in two different ways:

(a) potential as an electricity generation option;
(b) effect of direct solar heating and cooling applications on the electricity system demand requirements.

This paper addresses the first issue — the use of solar energy as a source for electricity generation. It is my purpose today to explore the role that the solar electric systems could play in meeting long-term electricity needs and to present a perspective on some of the utility system aspects of solar electric systems.

The goal of an unlimited and economic energy resource has intrigued man's ingenuity for many years. As early as 1913 a steam engine was powered by a solar-heated boiler. Similarly, geothermal heat was hoped to be such an inexhaustible energy resource since shortly after the advent of electricity itself. In 1904 the first geothermal plant was built at the Larderello field in Italy. With the discovery of the nuclear energy process in the early 1940s, the fission reactor was deemed this unlimited supply. Fusion systems were conceived shortly thereafter. Currently nuclear fission energy is the only one of this group that can be considered commercially available on a large-scale basis. Solar electric systems are now receiving much development attention, but, except for special cases, face substantial technical limitations which must be overcome in order to meet competitive economic goals.

It is natural to ask why it takes so long to achieve the commercialization of new energy concepts. The answer is complex, and it is important that solar scientists and engineers understand the problems. It cannot be assumed that solar electric research and development can bypass any of the pitfalls experienced by the other options, but perhaps careful attention to the factors influencing the transition from the laboratory to useful electricity supply may help smooth the way.

---

*Presented at the AAAS Annual Meeting, Denver, Colorado, 20-25 February 1977.

R & D Phases of Advanced
Generation Options

The pursuit of advanced power generation options requires decades of work and dedication. The typical development phases and their timing, as projected for some generation alternatives, are shown in Fig. 1. Programs must progress through scientific, engineering, and commercial feasibility before substantial utility integration and significant use are possible. Light water reactors are currently in the significant use stage and the fast breeder is proceeding through the engineering feasibility stage. The timing shown in Fig. 1 is characteristic of a peacetime rate of national resource application determined by such factors as technical uncertainties, supporting engineering tests, construction times, industrial manufacturing development, administrative and institutional mechanisms, and the national conception of competitive benefits. It has been possible under survival crisis conditions to speed this process by accepting very high risks of failure, by-passing the orderly development phases, and disregarding economic costs — as was demonstrated in World War II. However, in regard to future electricity supply, there appears to be no foreseeable need for such an extreme national effort, and thus the timing shown in Fig. 1 is probably the best that can be expected.

Solar-thermal electric research and development is now in the initial phase of demonstrating engineering feasibility. If all goes well, before the end of the century it may show its commercial attractiveness. It is important to remember that a time window exists for the introduction of each generation option. If the new technology is not ready at the proper time to replace competitive matured technologies, a new technology with marginal gains will not achieve acceptance in the market-place of end use on operating systems. The probable timing of a successful solar-thermal option shows its potential utility integration phase between the fission breeder and the fusion systems. The window for its application is therefore bracketed.

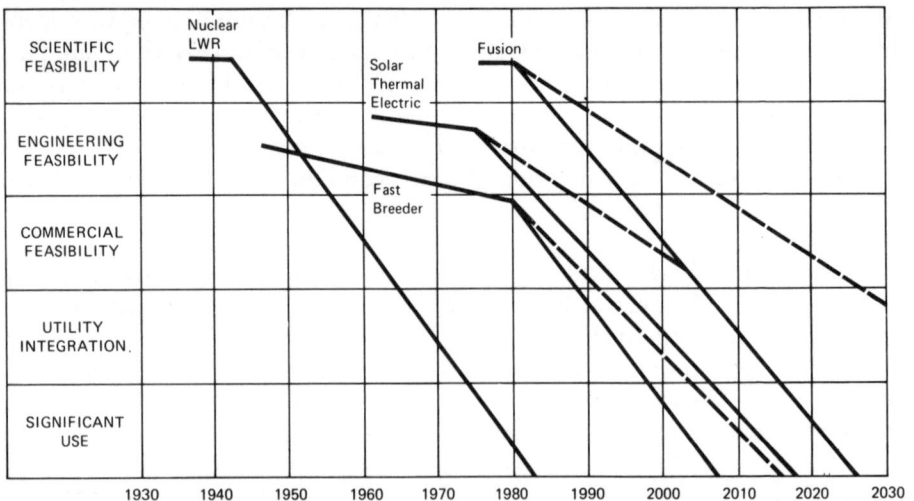

Fig. 1. Development phases for future power options.

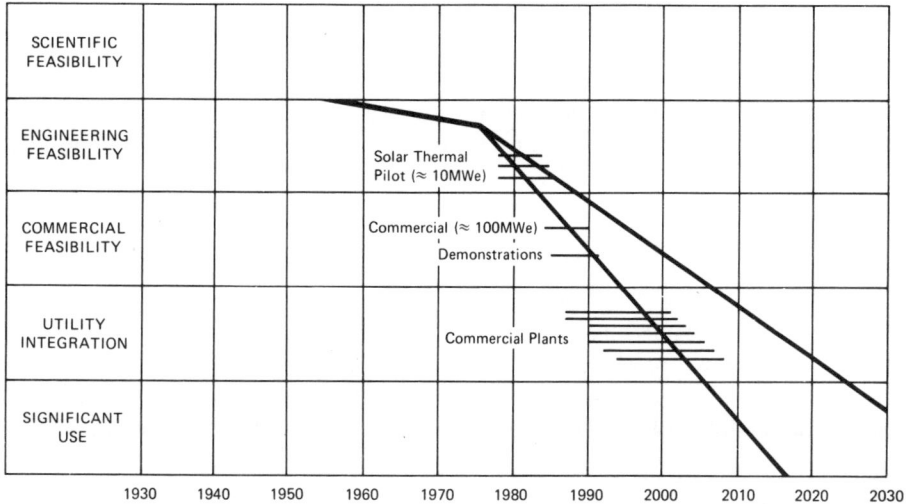

Fig. 2. Development phases of solar electric systems.

Figure 2 shows in more detail the status of the solar energy program in the conceptual framework of the development phases. ERDA has recently indicated its intent to proceed with a 10-MWe capacity solar generating unit. Other solar-thermal pilot installations of about the same rating are being considered. As design, construction, and operating experience is gained on these pilot plants, sufficient confidence may exist to initiate the construction of a 100-MWe demonstration plant that will be operational some time about 1990. For a substantial effect on the electricity generating needs at the end of the century, it would be necessary that all of the commercial solar generating units be constructed in the last decade of this century. EPRI projections of the electricity that could be generated from solar energy in the year 2000 have been consistently lower than other forecasts. We have projected that perhaps solar electric systems could contribute about 1% of the total electricity usage at the end of the century. To achieve this objective requires the demonstration of a cost-effective option prior to 1990 and the construction of 250 100-MWe solar generating units in about 10 years. It is our present belief that an introduction rate of this magnitude is an aggressive and perhaps optimistic goal. Nevertheless, this framework can be used as a base point in examining what research and development must be accomplished to achieve this goal and the utility system factors that would affect this introduction rate.

Factors Affecting the Commercial Integration of an Advanced Generation Technology

Opportunities for commercial introduction can arise as a result of diverse reasons, many of which are not apparent during the early research and development periods. Careful consideration must be given to the expected features of the competitive technologies at the time the new technology enters the market. For purposes of guiding R & D efforts, it would be imprudent to predict the outcomes of the many uncertainties that can affect the merits of the competing technologies several decades from now. If electricity demand is low, if coal costs remain low, or if uranium is plentiful, the opportunity for new generation systems could be small. Even after the achievement of technical success, new contenders cannot be certain to capture a major share of the power generation market. New electrical-generation alternatives do not create a new product to capture the consumer interest, but only the same end product, electricity. Emerging energy technologies are successful because they fill a user requirement at the time they are needed. The needs and the time window may not be fully perceptible 25 years before.

Nuclear power is such an example of a technology that became competitive in a manner not clearly foreseen at the time of its initial development. Early nuclear-power plant costs and fuel-cycle costs were high compared to coal generation costs. Today, although the nuclear-fuel-cycle and plant costs are higher than predicted, the coal system costs have experienced even greater increases, and nuclear is competitively attractive.

Since the solar-thermal generating options could experience a similarly changing situation, a thorough understanding of the criteria by which a utility will select new generation options is of utmost importance in judging its potential benefits and merits. These utility system criteria can be divided into four categories, as shown in Fig. 3: economic, resource availability, system capability and flexibility, and licensing. Key considerations affecting the ability of a generation alternative to compete for a share of the market are also indicated. For example, the reliability and maintenance aspects of a generation option expressed as "low forced outage rate" and "low planned outage rate" will have a profound effect on the economics of a new option. Each of these utility system criteria should be given consideration at all stages of the development of solar electric systems.

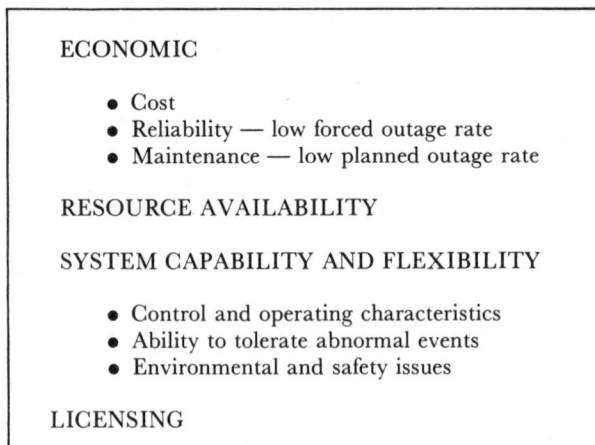

ECONOMIC

- Cost
- Reliability — low forced outage rate
- Maintenance — low planned outage rate

RESOURCE AVAILABILITY

SYSTEM CAPABILITY AND FLEXIBILITY

- Control and operating characteristics
- Ability to tolerate abnormal events
- Environmental and safety issues

LICENSING

Fig. 3. Utility system criteria for the selection of advanced generation options.

A specific solar-thermal electric concept will be selected to illustrate the type of questions that merit attention. In solar-thermal electric power plants, concentrated solar energy is converted to thermal energy and transferred to a working fluid for use in conventional Rankine or Brayton cycle turbine generators.

Two general classes of solar-thermal conversion systems are the central receiver and the distributed receiver. In a central receiver system, solar energy is optically focused by an array of two-axis tracking mirrors (heliostats) onto a central receiver or heat exchanger that is located on top of a large central tower. Solar energy concentration ratios from 1000 to 2000 are possible and temperatures of 1000–2000°F can be produced at the receiver for use in the turbines.

In distributed systems, solar energy is directly converted to thermal energy and transferred to a fluid at each individual collector. The fluid is then pumped through an extensive network of insolated pipes to a centrally located turbine generator plant. The collector and tracking options that have been investigated include two-axis tracking paraboloidal dish collectors, single-axis tracking parabolic trough collectors, stationary or seasonally adjusted flatplate collectors, fixed hemispheric mirrors with movable receivers. Because of the different concentration ratios of these various concepts, receiver temperatures range from 200 to 1500°F.

Recent analyses indicate that solar-thermal electric power plants will be most competitive when used to meet intermediate load requirements, displacing conventional power plants using scarce oil and gas. Also, their application will probably be limited to the southwestern United States, a region with high direct insolation. Current estimates indicate that the central receiver concept delivers the lower busbar energy costs of the alternatives considered. This is due to three factors:

(1) The higher receiver temperatures result in more efficient conversion of solar energy to electric power and smaller mirror areas per rated electric power output.

(2) The optical transmission of the solar radiation to the central receiver eliminates the high costs of insolated pipe required by the distributed system.

(3) The large central receiver uses a more efficient and lower-cost heat exchanger. (The multiple heat exchangers used by distributed collectors often require selective coatings and vacuum envelopes to efficiently absorb solar energy and to minimize thermal losses.)

The specific concept that was selected for more detailed consideration is shown in Fig. 4. This 100-MWe solar-thermal system is located in an area of high direct solar insolation in the southwest United States. The central receiver concept is of special interest to the Electric Power Research Institute, since it is a key part of its solar energy R & D program.

A 260-m central tower with thermal receiver is located in a field of flat solar reflectors covering a land area of approximately 1.3 km². Major energy conversion equipment is located in the tower. Two possible gas turbine thermal systems are being studied at the present time. One uses a high-temperature closed helium cycle

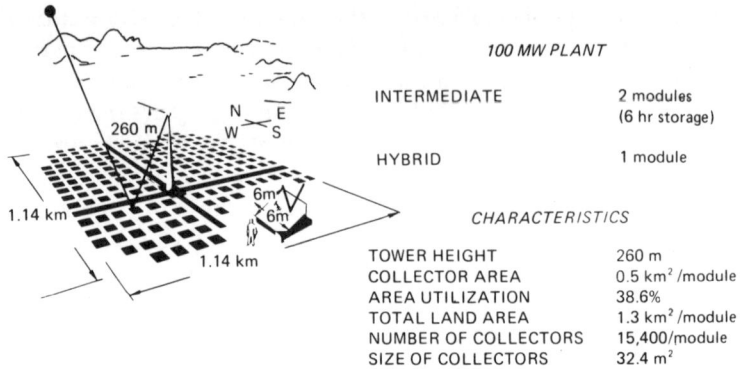

Fig. 4. Solar thermal conversion: central receiver concept.

and the other an open-cycle air system. Heat rejection in both systems can be accomplished without the need for cooling water.

Two alternative applications of the basic 100-MWe modules are indicated. These modules would be applied as follows:

(1) intermediate load application — two modules with 6 hours storage;
(2) hybrid system — one module with auxiliary oil-fired system.

Some of the specifications for the application of solar-thermal electric systems are shown in Fig. 5. These systems are capital intensive, and system lifetimes of 25 to 30 years must be achieved to approximate the economics of competitive systems.

CAPITAL INVESTMENT — 2 TO 3 TIMES COAL-FIRED STEAM UNITS

DEMONSTRATED 25–30 YEAR SYSTEM LIFETIMES

ENERGY STORAGE

REFLECTOR INSTALLED COST — $8 TO $10 PER SQUARE FOOT

APPLICATION

Southwest
Intermediate Load — 6 hours' Storage or Hybrid Mode
Capital Credit Probable

Fig. 5. Specifications; solar thermal conversion systems.

Economic Criteria

Our perception of the projected electricity costs from solar electric systems is that they will be substantially higher than that anticipated for other base-load options that will be available in the latter part of this century. The projected base-load costs, in 1976 dollars, for both the coal and nuclear options are approximately 40 mills/kWh. The major contributor to the cost of electricity generated from solar electric plants is the capital investment required to construct the unit. In addition, the variable nature of the solar energy source probably limits the number of hours of operation to that experienced by load following intermediate generation facilities.

The projected costs per kilowatt hour of a number of advanced generation options, including the solar thermal option, are shown in Fig. 6. For solar base-load operations with capacity factors of 0.6 or above, the most optimistic projections do not result in generation costs of 40 mills/kWh. However, oil-fired peaking and intermediate-load generation use could result in electricity costs in the 60 to 70 mills/kWh range. It is in the applications with capacity factors of about 0.4 that we see the initial introductions of solar electric systems. For example, Fig. 7 compares the cost per kilowatt hour as generated from the 100-MWe solar-thermal electric plant relative to electricity supplied by a combined cycle unit, using oil at varying prices. As the cost of oil rises, the solar-thermal unit becomes competitive. A breakeven could occur when the oil costs become about $30 per barrel.

These solar-thermal electricity costs were calculated using target assumptions for capital costs and an ideal location at Inyokern, Southern California. Six hours of storage are included in this concept. The comparison is made for 4000 full power hours of solar electric operation per year which should be achievable at this location. To supply a similar electrical load with the same solar electric plant plus a backup combustion turbine in a less desirable Northern California location would result in the upper cost curves. The average daily direct insolation in the latter case is about three-quarters of the Southern California example.

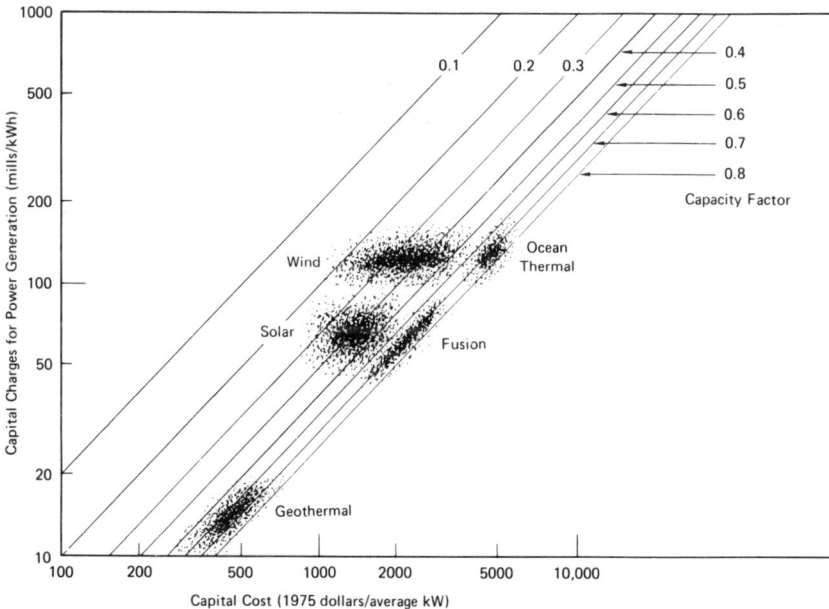

Fig. 6. Capital costs and capital charges for new electric power generation options.

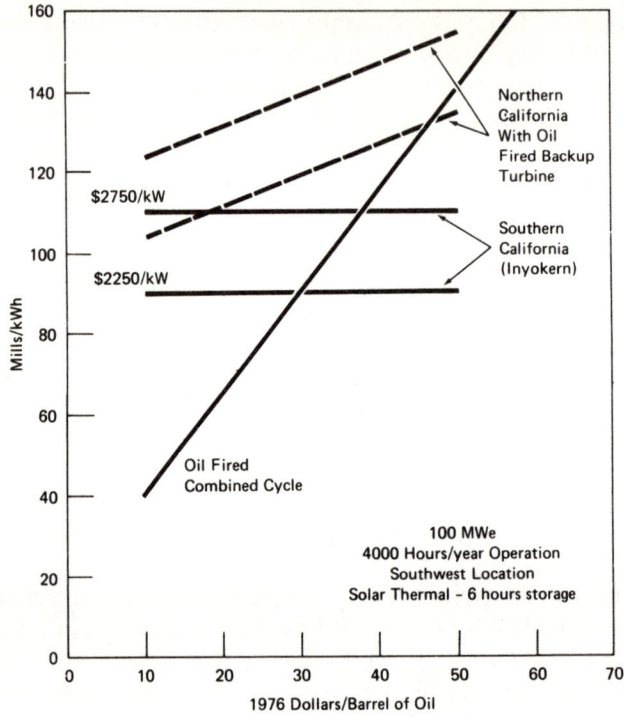

Fig. 7. Comparison of target electricity costs.

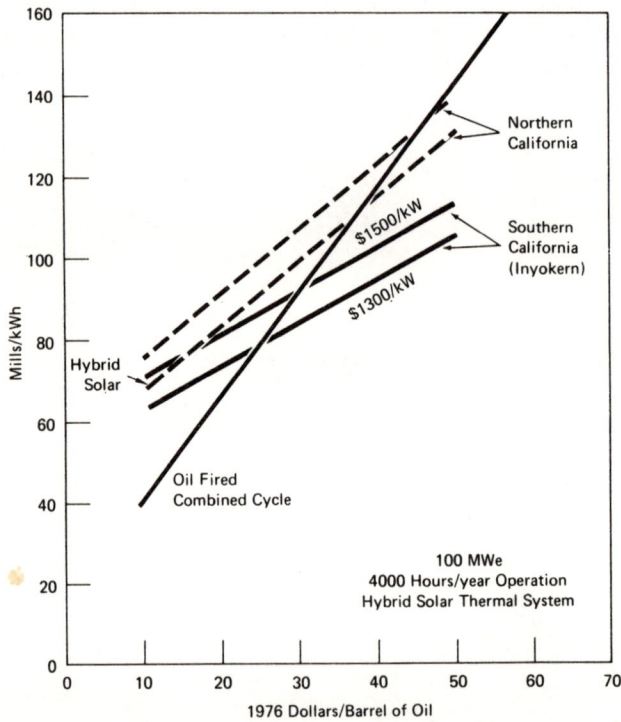

Fig. 8. Comparison of target electricity costs.

A similar analysis using the EPRI proposed hybrid concept results in the electricity costs shown in Fig. 8. In this case the breakeven cost for the Southern California location occurs at an oil cost of $25 per barrel and for a location with three-quarters the direct insolation of Inyokern at $40 per barrel.

Reliability

Most of the new generating options for the next century are conceived as having very low fuel costs but high capital investments. Systems in this category, such as the solar electric, must operate with as high a capacity factor as possible so as to minimize the cost of generated electricity. The relative economics shown on Figs. 7 and 8 was for 4000 hours per year of operation. Forced or unplanned shutdowns during the daytime period are particularly undesirable. Deficiencies that could cause long shutdowns are usually the most significant contributors to the forced outage rates. Mature nuclear and coal plants with scrubbers are now operating with unplanned forced shutdown rates of 15% and 18% respectively. Aggressive development programs are underway at EPRI to decrease these percentages substantially, perhaps to less than 10%. It should be possible to design solar-thermal systems with low forced outage rates such as 10%. Nevertheless, a number of design points could probably benefit by special attention. The high operating temperatures of the referenced solar thermal systems and the temperature cycles arising from the need for a daily startup of the plant could present a formidable problem. Experience with power-generating units shows that frequent shutdowns and startups cause more equipment problems than continuous operation. Systems requiring a complete cooldown of the thermal conversion system could tend to increase the forced outage rate. Design questions relative to the merits of maintaining the thermal system hot during the night deserve some attention.

Besides the issues related to the reliability of the power-conversion equipment, attention must be given to the forced outage rate that must be assigned to the field of solar mirrors. The effects of natural phenomena that occur in areas of high solar insolation must be accommodated. Dust storms are an example of the type of event that may result in plant outage time. Attention must also be given to more serious storms with wind velocities of perhaps 125 miles/hour. While the plant need not operate during such a rare condition, it would be desirable that long shutdowns would not occur as a result of such an event. In one respect, solar energy systems have the potential advantage that much of the maintenance can be performed at night. As a result, the inherent forced outage rate can be reduced by attention to those features that would reduce the need for shutdowns exceeding 10 hours. Design features allowing rapid repair of the equipment will be an important contributor to achieving a competitive system.

The design of the solar mirror field should also consider ease of maintenance. Access for equipment to clean the mirrors is an obvious requirement. Sufficient space to easily repair or replace a mirror would also be essential. With 15,000 mirrors in the field, attention should be given to achieving low probabilities of failures of individual units. Of course, deficiencies in single mirrors would not significantly affect the plant output, and maintenance on individual mirror assemblies should be possible during operation.

Resource Availability

Resource questions related to the advanced options are of significant impor-tance. These issues relate to both the energy resource and the water requirements.

While solar energy is an inexhaustible source, solar electric generating units do have resource availability problems and are subject to a range of operating conditions, uncertainties of solar insolation, diffusion by cloud cover, and water resources.

For example, Fig. 9 shows the solar energy insolation in various parts of the United States, in thermal megawatts per square mile. This data represents the energy equivalent of the total radiation impinging a horizontal flat plate collector. A factor of less than 2 exists between the lowest region and the highest region indicated on the chart. Nevertheless, a number of points must be considered in interpreting these energy levels. For example, the probability of cloud cover and the duration of such conditions varies significantly throughout the country. For example, Fig. 10 shows the isopleths of direct solar radiation in the United States. While the southwest section of the country enjoys high daily direct solar radiation as a result of infrequent cloud cover, the remainder of the country receives much indirect radiation. For example, Northern California receives less than one-half the direct solar radiation of Southern California or Arizona. As a result, solar-thermal electric systems that require direct insolation would be initially applied in the southwest, where the probability of cloud cover would be low.

The efficiency of utilization of the available direct insolation represents another important aspect of solar energy. The energy losses in the referenced hybrid solar-thermal design are illustrated in Fig. 11. The design study for this system accounted for tracking losses, light transmission and reflection losses, thermal losses at the receiver and the turbine generator cycle losses. An overall efficiency of 18% is believed to be achievable with this design.

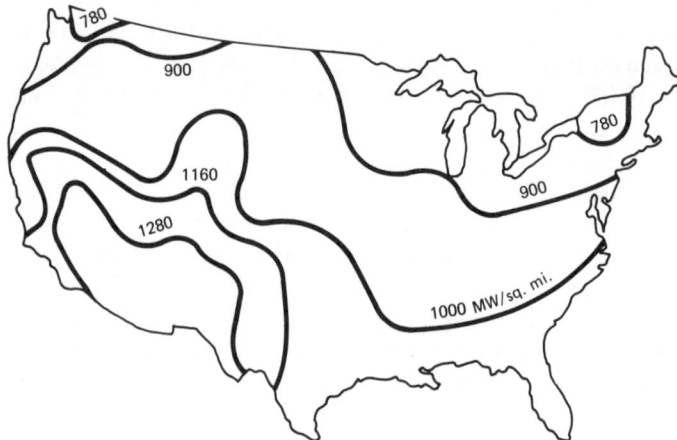

NOTE: Earth's surface, clear day, sun overhead,
insolation = 2600 MW/square mile

Fig. 9. Average daytime (12 hour) solar energy insolation (megawatts per square mile).

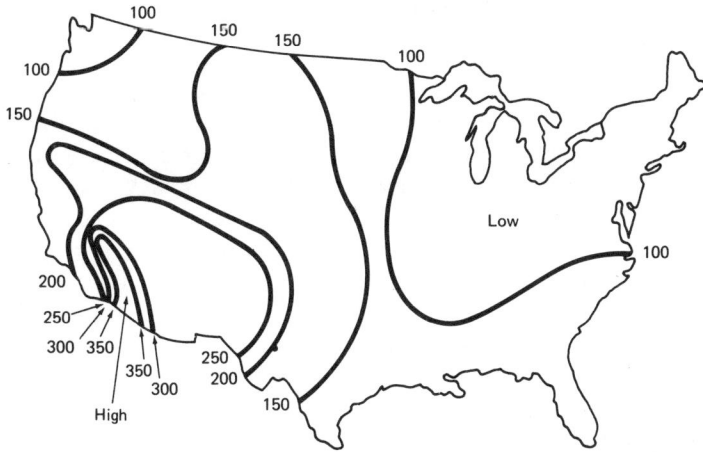

NOTE: Isopleths of mean daily direct solar radiation (Langleys).

Fig. 10. Choice of siting area.

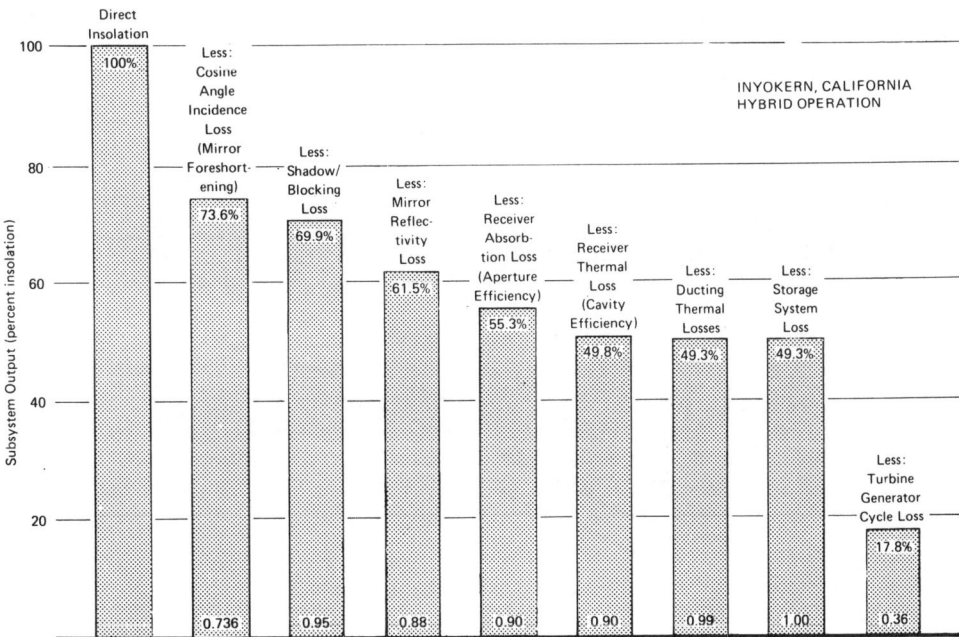

Fig. 11. Solar thermal system performance.

System Flexibility and Compatibility

The characteristics of electricity-generating options must be compatible with existing electrical systems. Flexibility of operation under a wide range of conditions is a highly desirable characteristic. Solar energy has the problem of being a variable solar source of energy, hence a backup energy electric system may be required. The application to specific utility systems will therefore require a substantial amount of data on the expected solar insolation at the specific site. In most applications of solar-thermal systems, backup generation capability will be required to maintain the overall utility system reliability. With the hybrid unit this backup capability would be designed into the solar heat cycle. For the solar thermal system with 6 hours' storage, this backup capacity would probably be accomplished with added combustion turbines. Such a situation could increase the capital expenditures of a utility to the point that the solar system would be unattractive. The economic application of solar systems should be enhanced in utilities whose peak power requirement occurs on sunny days which are coincident with the time the solar electric unit could operate with high availability. The results of an interesting evaluation of the backup required as the solar capacity increases are shown in Fig. 12. Values for a generating unit at Inyokern and at Santa Maria, both in California, are shown. The former is a desert location and the latter receives more cloud cover. If the solar capacity represented 20% of the total electrical system, backup equal to 40% of the solar capacity would be required at Inyokern and 50% at Santa Maria. If diversity is obtained by location of generating units at both sites, the backup capacity can be reduced to 32% of the solar capacity.

The total system performance of the design should be considered as the 10-MWe pilot plants are being developed. Typical examples of design criteria that must be considered are:

(1) Ability of the plant to drop load completely without excessive thermal stresses on parts of the system.
(2) Ability to supply the needed electricity when a condition of partial cloud coverage exists.
(3) Ability to start up rapidly so that long heat-up time is not required.
(4) Ability to operate during partial cloud coverage and other events that will occur at reasonable frequency.

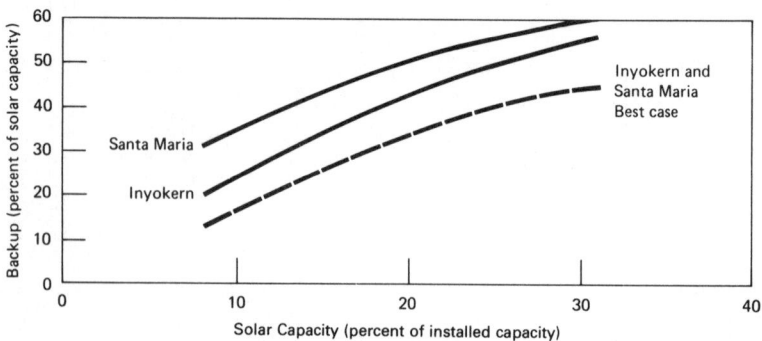

Fig. 12. Dependence of required backup capacity on solar capacity.

Licensing

While the licensing of solar electric systems may eliminate some of the issues associated with current generating options, the question of land use is an important one. The issue of the land needed for the solar collector field is well known, but transmission system right-of-ways also require much land. Concepts assessing the transmission of electricity over long distances must consider this factor.

Another potential licensing problem is that of water availability. Areas with high solar insolation usually have limited water sources. The water that is available is likely to be used for other important purposes and may not support a solar-thermal electric-generating unit. For this reason the EPRI prime interest is directed toward the Brayton cycle systems that could be based with dry (air) cooling for heat rejection.

Potential Availability of Solar
Electric Systems

Whether solar systems will be available to compete significantly with other options in the first quarter of the next century remains an important question. As indicated in Fig. 2, development schedules show engineering feasibility in 1983 with the operation of the central tower 10-MWe demonstration. This step must be followed by a 100-MWe prototype demonstration. It would be overly optimistic to assume that a commercial unit following the prototype could be built on a risk basis by the utilities and the manufacturers in less than 6 years. During this same time period, 1976 to 1990, the price of oil for utility use could increase to the breakeven values shown on Figs. 7 and 8. Alternatively, coal liquefaction processes could be developed by 1990 to supply a substantial part of the utilities' oil needs. The possibility exists that the costs of liquids from coal could also be high enough to enhance the competitiveness of solar-thermal electric systems. The comparisons at that time must also include competition from higher efficiency systems, such as fuel cells and higher temperature combustion turbines. Optimistically, an economically competitive solar-thermal plant could operate in about 1995.

To gain a perception of one of the main contributors to the time required to achieve economically beneficial application, consideration is directed to the investment requirements of the phases of an energy R & D program as shown in Fig. 13. Subsequent phases entail increased development costs, investment, and the associated increased risk.

To approach the time scale of some 30 to 40 years, the early phases of the research and development must be federally supported. Large investments by the manufacturers or the electric utilities cannot be justified, since present discounted values of long-range developments are too small. Therefore, the scientific research phases of long-range developments should be primarily federally supported, the later engineering and demonstration phases should receive substantial support by both the government and the utilities, and the commercial phases should be the prime responsibility of the manufacturers and the utilities.

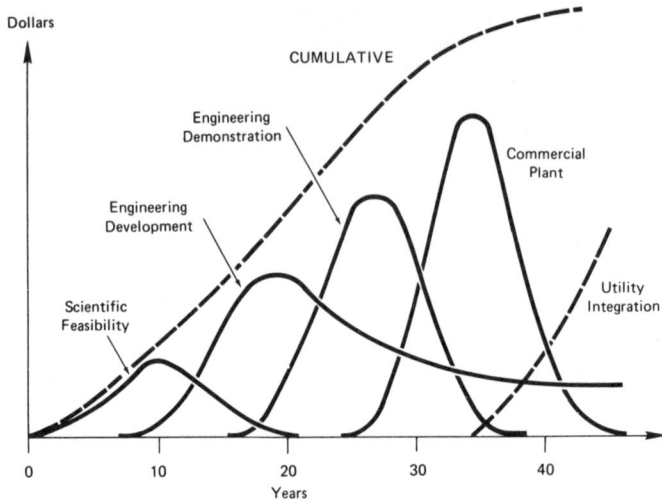

Fig. 13. Phases of R & D.

The relative costs of the R & D phases shown in Fig. 13 illustrate a number of important points. During the scientific feasibility phase, the research and development costs are low relative to the engineering demonstration and commercial plant stage. The opportunity should exist to examine many concepts and determine the proper one for the later stages. Solar-thermal electric systems are in the engineering development phases while some of the other solar concepts are in the feasibility stage.

The cost of a venture such as solar-thermal generation motivates the cautious developer toward a sequential research and development approach. One phase is completed before the next one is initiated. Parallel and competitive approaches are discouraged because of the large investment required. As a result of this, it is difficult to accelerate the commercialization of an advanced system.

Growth from the point of commercial availability will depend on the utilities' perception of the competitive advantage of solar-thermal systems, the capital available for generation expansion, and the rate at which manufacturing facilities can be built.

To obtain an understanding of the rate at which a new technology can be introduced following the first commercial demonstration, it is of value to examine the experience with the LWR plants. The Shippingport nuclear-reactor plant represents the engineering demonstration of the light-water-reactor technology. Before this demonstration plant began operation, the commercial integration phase had begun. Orders had been placed for the Yankee, Dresden I, and Indian Point I nuclear units, but additional orders were limited until these commercial units were operated. As a result of the initial operating experience, utility confidence in the commercial competitiveness of the concept increased. Light-water nuclear-power plants were built by different utilities and manufacturers, which contributed to the growth of a competitive market. Figure 14 shows this increase of operating light-water-reactor plants versus time. It should be expected that a similar process will occur with any advanced generation concept, i.e. following an engineering demonstration there will be:

(1) A period of testing the commercial feasibility of the new technology, during which two or three units will be built and operated. This will allow a number of utilities, manufacturers, and architect-engineers to gain needed experience.

(2) A period characterized by a rapid increase in orders and hence in the number of units that achieve operating status. This is the maturing stage of the technology. Design, construction, and operation of a substantial number of generating units will be in progress so that there will be a need to build up design staff, manufacturing facilities, equipment-test facilities, construction expertise, labor forces and operating personnel. It will be during this phase of the utility integration that many of the time-dependent deficiencies of a maturing technology become evident. At this time the number of units under construction and in the initial operating phase is increasing most rapidly, and the requirement for research and development support will probably be at its peak.

(3) The final phase in utility integration will be achieved when the technology has fully matured to the point that the operating characteristics are close to the best that can be achieved.

From this example, it can be seen that about 20 years are required to progress from a commercial feasibility stage to that of significant use. It must be kept in mind that the project completion time during the early introduction phases of the LWR was about 4 years while advanced systems will probably require 6 to 10 years.

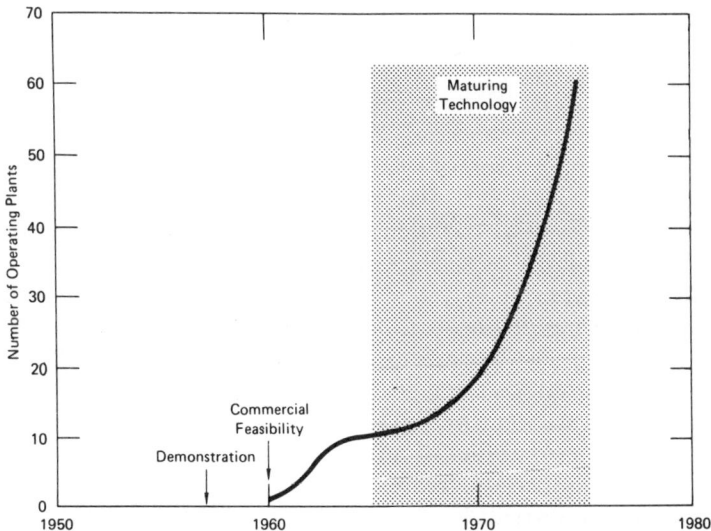

Fig. 14. Utility integration phase — light water reactors.

A perspective of the impact of these effects on the introduction of solar systems is shown in Fig. 15. The operation of the first commercial solar-thermal plant in 1985 is postulated as being similar to that experienced with the first commercial light water reactors in 1960. The increase of operating solar plants shown in the figure follows the number of light water reactors built between 1960 and 1975. The plant orders are shown by the dashed curve. With a 6-year lead time, some eleven plants would be ordered at the time the first commercial plant begins operation. This optimistic rate of introduction results in twenty solar-thermal electric units operating in the year 2000. If these plants are of 100 MWe capacity, only 2 GWe of capacity of solar energy is possible in the year 2000. Even if one assumed only a 3-year time span between commitment and operation, only about thirty-five generating units could be built.

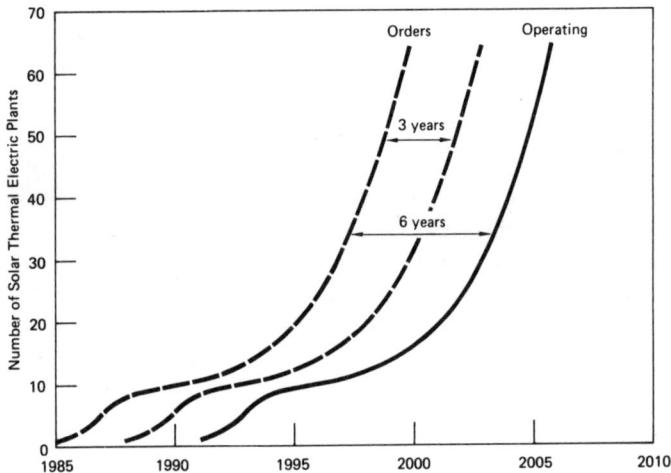

Fig. 15. Potential utility integration.

Summary

The substantial application of solar-thermal systems during this century is a challenge. Capital costs of the solar systems must be low and the initial plants must generate the essential confidence in the operation of the unit and future economics of the unit.

The most crucial step, today, in solar-thermal electric research and development is the engineering demonstration of some demonstration units of approximately 10 MWe each. As the various concepts are being considered, it is important to evaluate the engineering features and utility system compatibility as described in this paper. Some design features could have characteristics that will greatly increase the probability of adoption by the industry.

While this paper addressed some of the issues related to the application of solar-thermal electric systems, early attention to utility integration criteria should speed the demonstration of the competitiveness and operating features of any of the solar electric options.

4

*Nuclear Power and Weapons Proliferation – The Thin Link**

"Proliferation" — the current shorthand buzzword which describes the potential international spread of the production capability for nuclear weapons — has been a matter of national concern in the US since the closing days of World War II. But this recently popularized concern now appears to be causing a reversal of a quarter century of US policy regarding the best means of preventing proliferation. Stimulated by the last presidential campaign, the US has been moving toward prohibiting, or severely restricting, domestic use of the civilian fuel cycle including plutonium reprocessing and postponing, consequently, the US breeder reactor option. This is being advocated on the ground that if the US foregoes civilian reprocessing and the use of plutonium, and delays the breeder, other countries — energy-hungry though they may be — will voluntarily deprive themselves of the full benefit of nuclear energy to follow our "moral leadership".

Thus the US would in effect be saying to non-weapons countries, including nearly all of those aspiring to a higher standard of living through more abundant energy, "We and the other great nuclear powers already have our weapons, but to prevent the danger of further spread of nuclear weapons we ask you to follow us in giving up the most economic use of civilian nuclear power". The contradiction of such a posture is too clear to require further comment. Unfortunately, it is a demonstration to non-weapons countries that having weapons can be an economic as well as a military advantage. This has already been noted publicly by senior spokesmen of several countries.

Proliferation is a serious world-wide concern. If any nation, large or small, could threaten its neighbors with massive and rapid destruction, international relations would become more frightening. However, it should be recognized that with the technical capability so widespread, proliferation has been historically very much less than would be the case if the desire to have nuclear weapons was common. Clearly, to many nations nuclear weapons have only a marginal value — certainly less than that of a military air force which they all want and have. Perhaps history tells us that proliferation is likely to be judged by small nations as not in their self-interest. Of the 100 non-weapons countries which have signed the Non-Proliferation Treaty, about half already have the basic technology to produce weapons materials, and there is no indication that they have done so.

---

*Presented at the American Power Conference, 39th Annual Meeting, Chicago, Illinois, 19 April 1977.

The Administration's recent policy announcement on nuclear power has the effect of promoting LWRs and delaying both plutonium recycle and the breeder. The basic premise of this policy is that a combination of known coal reserves and uranium resources yet to be found permit our future energy needs to be met through the turn of the century without recycle or the breeder. The LWR fuel would pass through the reactor only once, and then be stored indefinitely. Because the uranium resource future is uncertain, many of us believe the insurance aspect of the breeder option (which requires recycle) should be fully developed now and subsequently used as required. All other advanced fuel cycles which give high utilization of uranium or thorium also require reprocessing. The administration's counter-argument is that closing the plutonium fuel cycle, as required for a fully developed breeder system, would place the US in the position now of endorsing plutonium recycle, and thus encourage the development in other nations of a possible channel for supplying weapons material. It is also the Administration contention that domestic pursuit of the breeder and recycle for US energy supply, and simultaneous discouragement of other nations, would create an unacceptable double standard — although such already exists in the weapons field. Many of us have a deep concern that the Administration's position on plutonium recycle and the breeder will be internationally counter-productive and actually stimulate proliferation, and may domestically damage our economy as well.

How real is the danger that reprocessing of civilian fuel would be used by nonweapons countries to obtain plutonium for weapons? A simple analysis of well-known facts shows that there are today no fewer than eight different ways available (Fig. 1) to produce weapons material — to *produce*, not steal. Among them, the route of commercial nuclear power using slightly enriched uranium fuel ranks eighth and low in desirability for a country that has made a political decision to establish a nuclear-weapons capability. It is the most expensive route (five to ten times more costly), requires the highest level of support technology and the broadest base of support industry, and takes the longest to install and to yield material (3 to 5 years longer).

By the 1980s, moreover, the world may have available several new, additional ways (Fig. 2) of producing weapons-grade fissionable material, adding even more routes to weapons capability. Several of these are likely to be easier, cheaper and

	Required		
	Cost	*Technology*	*Industry*
Research reactor	Small	Small	Small
Production reactor	Medium	Medium	Medium
Power reactor	Large	Large	Large
Diffusion cascade	Large	Large	Large
Centrifuge cascade	Medium	Medium	Medium
Aerodynamic jet cascade	Large	Medium	Large
Electromagnetic separation	Medium	Large	Medium
Accelerator	Medium	Medium	Medium

Fig. 1. Eight-fold ways available for weapon material production.

```
┌─────────────────────────────────────┐
│  U-235 SEPARATION                    │
│  • Laser                             │
│  • Chemical exchange                 │
│  • Jet membrane                      │
│                                      │
│  Pu or U-233 PRODUCTION              │
│  • Plasma fusion-fission             │
│  • Inertial fusion-fission           │
│      Laser implosion                 │
│      Electron beam implosion         │
│  • Accelerator (ING)                 │
└─────────────────────────────────────┘
```

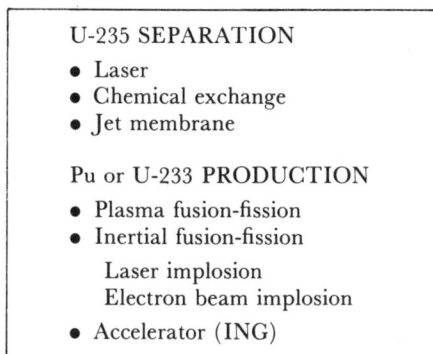

Fig. 2. Potential added routes to nuclear weapons (in the 1980s).

faster for a nation bent on bootstrapping itself to acquire weapons capability than is the route of nuclear fission power with reprocessing and recycling of fuel.

The question thus arises: if the dike is leaking in at least eight places, maybe more, why are some so desperately anxious to plug only one of the leaks? Why this effort to focus attention on an issue that does not go to the heart of the problem? Why not pick also on centrifuge or laser enrichment? The breeder and recycle are clearly essential for the eventual longevity of nuclear fission as a world-wide energy source. It has been suggested, therefore, that the focus on the nuclear-power route to proliferation has been stimulated by those with an intense ideologic goal of stopping civilian nuclear power. Its opponents have failed to achieve this end on the issue of health and safety, on the environmental issue, on the waste-storage issue, and on the plutonium-hazard issue. Perhaps they have now fixed on the proliferation issue as a likely one to gain public support in the US — deliberately shrouding the inherent contradiction that if a maverick foreign government should decide to establish its own weapons material production capability, the civilian nuclear power route would be among its least attractive choices.

On economic grounds chiefly, I am personally dismayed at the concept of selling reprocessing facilities to small nations. It takes a large number of nuclear power stations to supply the flow of fuel needed to make a present reprocessing plant cost-effective. Further, what would be done with the output of plutonium and uranium? If self-sufficiency is the reason for closing the fuel cycle, fuel-fabrication facilities and possibly enrichment would also be needed. The total investment would be several times that of the initial power plant alone. Under these circumstances, such an unwise investment should be discouraged by con-scientious supplier nations. How many auto manufacturers try to sell an oil refinery to their customers?

What is needed, of course, is an assured fuel-cycle system, preferably under international auspices, operated so as to optimize the use of our world's resources and to inspire confidence·in supply. I am disappointed that the supplier nations, and the US in particular, have not energetically proposed such a system, although it certainly must be under consideration. Clearly, this would make civilian nuclear power a negligible part of the proliferation alternatives. Compared to the Administration's present negative approach of deferring a program that is needed by most of the industrial world and eventually by the US, the internationalization of the fuel-cycle facilities would be a positive step toward establishing an enduring trust in an international fuel supply. Such a system would provide a congruence of economics, national self-interest, and safeguards against proliferation. It is not obvious why this course has not been urged more forcefully, as it provides a long-term viable solution to the world-wide use of nuclear energy.

Just as quixotic is the argument that by self-denial we can motivate poorer, less-developed countries to renounce optimum use of civilian nuclear fuel. Reprocessing technology has been set out in ever-growing detail and shared at international technical conferences since the US first reported on the design and operation of the Idaho Chemical Processing Plant in 1955 at the first US International Conference on Peaceful Uses of Atomic Energy in Geneva.

It must not be forgotten that the US actively pushed the Non-proliferation Treaty, by which 100 countries have pledged that they will not make nuclear weapons, *in return* for international cooperation and nuclear interdependence on civilian nuclear power. Now the US is proposing in effect to reinterpret unilaterally its own commitments under that treaty and to withhold the pledged cooperation on civilian nuclear power, thus stimulating other nations to establish nuclear independence. To do this in the hope of motivating other countries to couple renunciation of nuclear power with previously pledged abstention from nuclear weaponry is provocative to the disadvantaged nations involved.

At a conference in New York last month on International Commerce and Safeguards for Civil Nuclear Power, speakers from country after country explained that for them nuclear power is not a debatable option but an indispensable necessity, and that reprocessing is an essential.

Said a West German: US denial of reprocessing or breeder technology at home would create greater pressure on others to move even faster with their domestic reprocessing and breeder programs.

Said a French spokesman: Nuclear power is indispensable to developing countries, and France will provide and guarantee fuel-cycle services without any pressure on other countries to forego development of nuclear capability alone. For France, reprocessing is absolutely essential, and she is not even studying the option of storage of spent fuel. The imposition of new conditions on exports merely serves to create greater distrust among suppliers and importers, which becomes an impetus to proliferation.

The Spanish spokesman declared: There is not justification for creating a problem with such serious consequences without at the same time suggesting an urgent and immediate solution. Any delay in making a decision to this effect may cause irremediable damage to countries that cannot take part in these decisions.

Said the speaker from Japan: Japanese confidence in dealing with the US appears to have been shaken. Strategic considerations exceed economics as 60% of Japan's electricity is generated from imported oil; therefore, the breeder is an indispensable part of Japan's nuclear program, which is the only option available to protect its energy capability.

At the same meeting, Commissioner Kennedy of the US Nuclear Regulatory

Commission said he believed that restricting US exports or imposing restrictive criteria would destroy US influence in the world market and lead to loss of whatever control the US might have on weapons proliferation.

There is overwhelming logic against the change in long-established US policy.

(1) The world's long-term needs for energy — not only in resource-starved countries, but in a few decades also in nations now exporting oil — do not permit ineffective or uneconomic use of nuclear fuel resources. United States willingness to use its own nuclear resources ineffectively and/or uneconomically is hardly likely to persuade resource-short countries of the merit of our moral and ethical leadership and stimulate them to follow such patterns.

(2) Embargoes or stringent restrictions on civilian nuclear power cooperation and supply of materials and services by the US and other supplier countries will only accelerate further the nationalistic trend toward construction of independent indigenous-fuel-cycle capabilities — both enrichment and reprocessing — and thereby lose or reduce the likelihood that effective international controls and safeguards may be accepted.

(3) In any country, the civilian nuclear-power program is separable in timing, resources, and institutions from a program to obtain nuclear-weapons capability. Historically, the major nuclear-weapons powers all made weapons first and developed civilian power later.

(4) The existence of spent fuel from power reactors in any country does, to be sure, represent a potential for deriving low-grade weapons material. However, as already stated, if a nation decides to produce weapons material, there are at least seven other routes available not involving power reactors, that are much less costly, more rapid and flexible than civilian power plants as a source of such material.

(5) Denying a government access to civilian reprocessing does not erect a significant obstacle or delay in implementing a decision to produce weapons. Spent fuel from *either* research *or* power reactors can be reprocessed rapidly and with relative ease — and *especially* if it is done without the various commercial and legal constraints that apply to civilian reprocessing plants. With the "once-through" type of fuel cycle, diversion does not even disturb the productivity of the power cycle.

(6) To minimize the possibilities of proliferation of weapons as a by-product of civilian power, it is most desirable to place the sensitive parts of the civilian fuel cycle under internationally supported and cooperative safeguards. Control of the fuel cycle is the key to safeguards, not control of reactor operations. This means control of enrichment, spent-fuel storage, reprocessing, refabrication, and shipment for recycle.

(7) One of the most evident and increasing causes of international tensions is energy malnutrition, as evidenced by increasing oil imports, worsening foreign-exchange deficits, and the limitations on costs and productivity associated with them. Inhibiting the effective use of civilian nuclear-energy supply in countries that have limited energy options open to them can result in enhancement of their propensity for international conflict.

(8) The US, with its extensive coal resources and remaining oil reserves, may find it less constraining than other nations to delay further the effective full use of

nuclear resources including reprocessing and the breeder. But the costs of delay will include further erosion or loss of what remains of US influence on the course of world development of effective nuclear safeguards.

(9) Effective international cooperation on safeguard systems for spent-fuel storage and for reprocessing has strong world-wide mutual motivations for both internal anti-terrorist security and economic reasons. Strong and timely leadership by the US (and other nuclear supplier countries) can help bring such systems into being.

(10) There is no nuclear-fuel cycle that is inherently "diversion proof" once the relative ease of military-style chemical reprocessing by a government is recognized (although the possible variations in fuel cycles do differ in their requirements for manipulating materials).

(11) The so-called "tandem" fuel cycle, which some have urged as a "technical fix" for weapons proliferation, is counter-productive as it requires widespread construction of heavy-water reactors. This type of reactor has been widely used as a plutonium factory and as a tritium factory (needed for thermonuclear weapons) and can be run with natural uranium, easily available everywhere.

(12) Similarly, the U-233 cycle is of very limited use as a "technical fix", being logistically unavailable for at least 20 years. There now exist neither the U-233, nor the reactors, nor reprocessing plants required for producing the amounts that would be significant for energy purposes. Such facilities are uneconomic at present. They would require large subsidies to get started. In addition, U-233/U-238 reactors could still be used to make enough separable plutonium and U-233 to make weapons, so that the need for effective physical site safeguards would not have been diminished.

(13) As for the "terrorist" threat, effective safeguard systems have been in use for more than 30 years for military reprocessing nuclear weapons materials, and weapons themselves. In the US, even with civilian reprocessing the total inventory of separated reactor plutonium would not reach 10% of the already existing military quantity before year 2000, and need never exceed 20% of the already existing military stock if the plutonium is recycled. (It has long been a truism in the industry that the safest place to keep plutonium is to burn it in a power reactor.) Highly effective safeguard systems are also applicable to civilian-produced material.

It is certain that neither the US nor the world has the choice of severely limiting nuclear power, and that the reprocessing technology and breeder issues have only a thin and controllable link to proliferation. These facts, unpalatable to some though they may be, must be constructively addressed — politically and institutionally.

It is.a disservice to the people of the US — and the people of the world — to create the illusion that by putting restrictions on civilian nuclear power we have somehow solved the proliferation problem. Unintentional though it may be, such steps would undoubtedly be counter-productive to their stated objectives by creating resource conflicts, removing faith in the US umbrella to protect the welfare of its allies, and stimulating the expansion of indigenous nuclear capabilities abroad including enrichment and reprocessing.

Finally, one may well ask, who would benefit from such a policy? The oil-exporting nations, of course, and those nations which are continuing to develop all their nuclear power options without restriction. Our national policies continue to be flexible enough so that it is timely to urge a more comprehensive and realistic assessment of our planning options.

5

*Future Technological Options for the United States Electric Power Industry**

Current concerns with the worldwide availability of primary fuels have highlighted the urgent need for increased development of new technological energy options which use our resources more effectively. Historically, fossil fuels have been relatively abundant, easy to obtain at relatively modest cost, and have not presented large problems of availability. During the past few decades, however, the rapid increase in fuel demand has gradually revealed the economic and social costs of increasing the drain of these depletable resources. This situation has been further highlighted by a growth in the general awareness of the associated environmental impacts and a growing desire to diminish these impacts.

Proposals to cope with these energy issues are numerous. Some embrace the thought that conservation alone can resolve the problems or that energy-dependent life styles should be altered. Others look to futuristic technologies as the answer to coming energy needs. It is now generally recognized that there is no single easy answer to satisfactorily meet these needs. The rational approach is to undertake a series of activities which, first, try to selectively develop a variety of means of supplying low-cost energy; second, improve the efficiency of energy use; and third, accommodate life styles in a reasonable fashion to a somewhat more restricted energy diet. Even though every conceivable method of improving the efficiency of energy use and of supplementing energy resources may be developed, the foreseeable demands in the coming decades probably will be difficult to meet without some change in our energy-use patterns.

The search for an unlimited and economic energy resource has intrigued man's ingenuity for many years. As early as 1913, a steam engine was powered by a solar-heated boiler. Similarly, geothermal energy was considered a possible inexhaustible energy resource since shortly after the advent of electricity itself. In 1904 the first geothermal plant was built at the Larderello field in Italy. With the advent of an understanding of nuclear energy in the early 1940s, the fission reactor was deemed this unlimited supply. Fusion systems were conceived shortly thereafter.

Today all of these systems continue to be classified as advanced. Solar and

---

*Presented at the World Electrotechnical Congress, Moscow, USSR, 21-25 June 1977.

geothermal electric systems are receiving substantial attention but, except for special cases, require major improvements to meet the competitive economic barriers. Experimental breeder reactor systems have been operating for over 10 years in the US, France, England, and the USSR. The 250-MWe Phenix reactor in France has been operating successfully since July 1974 and plans are underway there for a 1200-MWe reactor.

The principal technology options which are currently receiving attention as future electricity sources are:

Geothermal
Solar Thermal
Solar Photovoltaic
Biomass
Fission Breeder
Nuclear Fusion
Wind
Ocean Thermal Gradients

The principal incentive for investigating these options is that in each case, the fuel cost is very low compared to anticipated fossil fuel and uranium prices. While these benefits may be offset by high capital costs, these systems could presumably provide protection against rapidly rising fuel prices or an unavailable supply, as has occurred with oil. Each of these options will be examined with respect to its potential incentives, the associated obstacles which must be overcome, and an assessment of its likely prospects.

Geothermal Energy

The three types of potentially usable geothermal resources are dry steam deposits (the most easily utilized), hot water or hydrothermal, and hot-rock systems.

The steam resources of the United States appear to be quite limited. This is the type of geothermal energy used in California at The Geysers, an area now producing 500 MWe. Today, The Geysers represents the total US geothermal generating capacity. Here the geothermal steam is used directly to power turbines. The steam is then vented to the air, which releases quantities of hydrogen sulfide, ammonia, and radon, which must be controlled. The high concentration of dissolved minerals in the steam causes corrosion problems for pipes, valves, and the turbine.

Geothermal hot water contains even more corrosive material than dry steam. Because of this, it is planned that the bulk of this water would be reinjected into the ground after passing through a heat exchanger capable of handling such a corrosive liquid. The use of a second (working) fluid to power the generating system, and need for the reinjection of water, would of course increase the cost of the system. The known US deposits of subterranean hot waters are in the West and in most cases far from population centers. The need for relatively long transmission lines would add a substantial cost penalty.

A more advanced technology is needed to obtain power from deposits of hot rock. Water must be pumped underground to be heated by the deposits of rock. It must then be piped to the surface and treated as geothermal hot water. The geologic feasibility of this system is not yet known, and the quantity of water required to make up underground losses may be excessive. These three types are illustrated in Fig. 1.

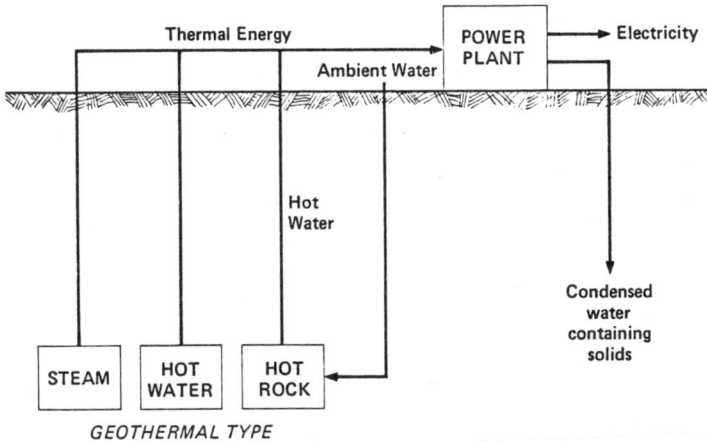

Fig. 1. Geothermal electric power generation.

All of these geothermal systems operate at a low thermal efficiency relative to conventional power plants, and as a result they produce more waste heat per unit of energy generated. A typical geothermal installation of 1000 MWe is expected to cover 15 to 30 square kilometers (roughly 10 times the area required for a 1000-MWe nuclear plant) and requires over 2000 wells.

The most optimistic hope in the US is that by the year 2000 roughly 40,000 MWe of geothermal capacity will be providing a low-cost energy source, or less than 3% of the electricity estimated for that time.

Solar Energy

Three different methods for the direct use of solar energy are being studied. The direct use of sunlight for space heating and cooling and for water heating is the most developed and is marginally economic in some areas. This will not be discussed here because it does not involve the generation of electric power. The other two technologies do produce electricity; these are solar thermal conversion and photovoltaic conversion.

Solar-thermal Electricity Production

Solar power is a large nondepletable source of energy which appears to be free from concerns over public health and safety. Additionally, the primary energy input is free.

The equipment necessary to capture sunlight and convert it to electricity is not free, however. This is one of the principal obstacles to its development. The diffuse nature of solar energy requires that a large area (roughly 1.3 to 2.6 km² for a 100-MWe central receiver plant) be covered with devices capable of collecting and focusing the available solar energy. The capital cost of a solar-thermal plant is now estimated to be three to five times that of a conventional fossil-fuel plant of the same capacity. Operating costs will be lower because no fuel is required, but this advantage could be eroded if the operating and maintenance costs of the system are high.

The limitations imposed by the intermittent nature of the source create another serious obstacle to widespread utilization of solar-thermal systems. Since solar energy is tied directly to the diurnal cycle, a substantial energy-storage system must complement any solar-thermal facility if it is to reliably provide dependable service through the evening peak-demand hour. The energy-storage system designed to complement solar thermal conversion plants must provide about 6 hours of output, which will provide for a 12-hour daily operation; COE (coefficient of performance) = 50%. Work is in progress on such facilities, but even when they are available they will add substantial cost.

Assuming a plant conversion efficiency of 15% of the overall insolation of the mirror area, about 1.3 to 2.6 km² of land is needed for a 100-MWe power plant. A 100-MWe plant would require 15,000-20,000 individually aimed 6 × 6-m mirrors and one or two central receiver towers, each 260 m high (see Fig. 2).

During a prolonged solar outage (several days of cloudy weather, for instance) reserve capacity, in the form of a conventional standby plant, may be needed to insure uninterrupted service. Thus, solar systems combined with conventional reserve capacity could substantially increase the capital intensive nature of the electrical generating capacity.

The best locations for solar-thermal conversion power plants are in areas with a large number of cloudless days and an abundance of flat, cheap land. This suggests the desert areas of the Southwest United States.

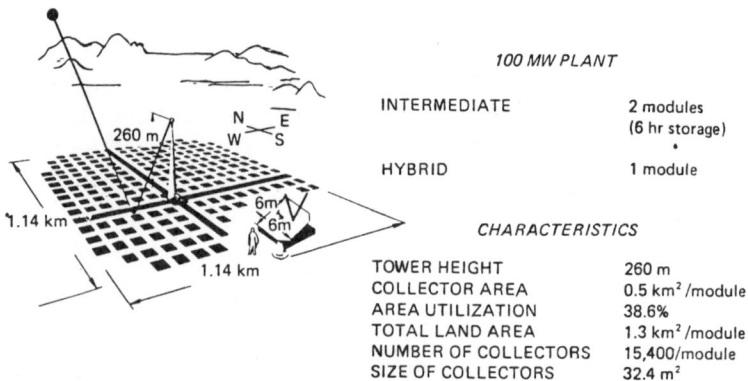

Fig. 2. Solar thermal conversion (central receiver concept).

Unfortunately, the supplies of cooling water in these areas are quite limited. The EPRI-proposed Brayton-cycle central-receiver concept plant uses air cooling, which eliminates this problem. This plant concept requires materials capable of long-term operation at very high temperatures.

A 10-MWe demonstration facility is planned for operation at Barstow, California, in the early 1980s. If it proves successful, the EPRI optimistic estimate is that solar-thermal conversion could provide about 20,000-40,000 MWe by the year 2000. This would provide less than 3% of the electricity needs at that time.

Photovoltaic Conversion

Solid-state solar cells may be used to convert light directly to electricity. The physical constraint of a large land area is slightly larger than that for solar-thermal conversion, yet photovoltaic conversion offers two additional benefits: for nonconcentrating systems, cooling water is not required because the light is converted directly into electricity, and diffuse sunlight, as might be present on a cloudy day, can be used. This would permit operation on cloudy days at about 20% of full sunlight capacity.

Counterbalancing these benefits are the low operating efficiencies of the cells (typically less than 15%) and the high costs of production. Reductions in cost by a factor of 100 to 200 are needed before this technology could be competitive with other means of electricity production. In any area in which large-scale photovoltaic power might be considered, solar-thermal conversion will probably be more economical. For this reason, photovoltaic conversion is not expected to make a significant impact this century, unless a scientific discovery gives a new low-cost and efficient photovoltaic cell.

Because of their intermittent nature, solar electric systems, even with energy storage of reasonable size, would be used for intermediate load applications rather than for base loads, so they will always complement coal and nuclear stations, rather than be a substitute for them.

Biomass

Wood is the oldest source of renewable fuel. As a result of the diminishing stock of fossil fuels, the establishment of plantations has been proposed to produce fuel for baseload power plants, as a means of tapping solar energy.

The plants that have been suggested include crops of sorghum, sugar cane, and sycamore trees, and the water crops of kelp and water hyacinth.

The obstacle to such programs stems from land requirements and harvesting costs. For continuous operation, a sycamore-fueled system requires 740 km² for a 100-MWe plant. The sugar cane or sorghum systems need about 130 km² per 100 MWe. These areas would compete with agriculture for available land, water and fertilizer, and for the plantation owners to make a return on investment equivalent to that from other crops, the fuel costs would be excessive. Unfortunately, a fuel crop represents the lowest value usage of a plant resource. Wood is more valuable as lumber or for paper production than simply as a heat source. Such plantations located on areas of marginal crop production are not as attractive as solar-thermal conversion, which requires much less area and can be built on land unsuited for agriculture. Biomass fuels, therefore, are not expected to make a significant contribution to electricity production in the foreseeable future, although some analysts forecast the use of biomass fuels to produce methanol as a gasoline substitute.

While the economics of energy plantations seem unattractive, by-product fuels such as farm and municipal waste will be used in the future. Farm wastes have been and will be used at the farm for low-grade heat for crop drying and other purposes. The cost of collecting and transporting farm wastes makes the direct use at the farm the most attractive and efficient way to use the fuel.

Municipal wastes, which must be collected anyway, can be used to fuel electric power plants. The total energy content of garbage produced in the US is roughly equal to 3% of present total energy consumption. Seventy percent of this municipal waste energy is in paper and plastics, and it may make more sense to recycle these materials directly than to burn them. While municipal waste may be significant in some urban areas, the low population density of rural areas makes the use of a substantial part of this waste impractical. The limited nature of this resource will limit its contribution to 2% or less of total US supply.

Fission Breeder

A breeder reactor resembles the conventional water reactor in most engineering-performance aspects, differing primarily in the fuel-element composition and the external fuel-cycle system.

The distinctions between the various reactor types are as follows:

(1) *Burner reactors* use highly enriched fuel (almost pure U-235) and produce no fuel internally. The pressurized light-water reactors used in nuclear submarines are typical of this type.

(2) *Converter reactors* require a continuing feed of fissile U-235 and fertile U-238 or Th-232 and produce fissile material (Pu-239 or U-233) at a rate slower than it is being consumed. Examples include the light-water-reactor system, which burns U-235 and generates plutonium, the heavy-water natural uranium reactor, and the high-temperature gas-cooled reactor, which requires a continuous supply of fresh U-235 for its operation, supplemented by self-generated U-233.

(3) *Thermal breeder systems* are based on the Th-232/U-233 cycle; however, they must be initially fueled with U-235. After a number of years they can attain an equilibrium condition where they produce U-233 at the same or slightly greater rate than they consume it. The light-water-breeder reactor is an example of such a thermal-breeder system.

(4) *Fast-breeder reactor* systems are fueled with plutonium and produce substantially more fissile material than they consume. The liquid-metal fast-breeder reactor (LMFBR) and the gas-cooled fast-breeder reactor are examples of this type of system. This excess fissile material can provide the first core inventory for a subsequent plant.

The liquid-metal fast-breeder reactor is close to economic development; there has been substantial international progress towards its development. In economic operation, an LMFBR will produce 50 to 80 times as much energy per kilogram of natural uranium as will the conventional water reactors. After the initial core loading, subsequent fuel addition is very small, and therefore the fuel costs are a negligible part of the cost of electricity produced by an LMFBR. This provides a significant incentive for development of a commercial reactor. Several decades of environmental experience with LMFBRs is available to provide a basis for the construction of commercial size plants (the first nuclear-produced electricity came from a breeder, EBR 1, in 1951). Because of the engineering similarity of LMFBRs to conventional reactors, much of the experience gained by the utilities in the operation of conventional nuclear plants can transfer to the operation of breeders.

The environmental effects from the normal operation of an LMFBR are expected to be similar to those due to water reactors. The waste-heat release from an LMFBR will be significantly lower than from a comparably sized water reactor due to the higher thermal efficiency (40% versus 33%). The land-use requirements for the plant and fuel-cycle facilities should be comparable, but the land required for the mining of uranium for use in light-water reactors would be saved by LMFBRs which could operate for centuries from present stockpiles of depleted uranium.

The experience gained through the operation of pilot and demonstration facilities (Phenix in particular) indicates that a breeder reactor can meet the performance criteria necessary for utility operation. High-capacity factors have been achieved with these reactors and in general they have performed as expected. All of this constitutes sufficient incentive to develop breeder reactors. The position of the breeder is unique — it is the only method currently available to provide essentially unlimited energy at costs which are projected to be competitive with conventional sources now in use.

The obstacles to early commercialization of a breeder reactor are not primarily technical, but rather stem from the problems faced by nuclear energy in general and particularly those social and political problems associated with management of the plutonium fuel cycle.

While there seems to be general agreement that an LMFBR will be required to meet the same stringent safety criteria required of LWRs, there is some uncertainty as yet as to what this means in terms of the engineering of plant-safety features. Public concern with the transport and handling of plutonium has focused on the breeder because the most economic breeder will be plutonium fueled. The plutonium-handling requirements for breeder reactors will be several times that for water reactors, although it seems quite likely that the approach to this problem developed for plutonium recycle in light-water reactors will be satisfactory for the breeder. The final uncertainty concerns fuel-recycle facilities. Only when it is clear that installations will be built to reprocess spent fuel and fabricate new fuel will the breeder be attractive, because the chief benefit of the breeder stems from its internal production of fuel.

Within the US there is as yet no national policy for early commercialization of a breeder reactor. The first French breeder reactor started about 15 years after the first American LMFBR, yet the French plan to have a commercial unit on line in the early 1980s.

The prospects for development of the breeder reactor in the US depend in part on policies which will be established by the new administration and Congress. Current plans call for the operation of the Fast Flux Test Facility (a 400 MWt LMFBR) in the late 1970s, the operation of the Clinch River Breeder Reactor (a 380-MWe LMFBR) in the early 1980s, and the first commercial operation in the early 1990s. If this progression occurs on schedule, some utility LMFBR capacity could be in use by the year 2000.

Nuclear Fusion

The theoretical incentives for harnessing nuclear fusion are substantial; it uses another new domestic fuel supply (lithium), its fuel cost will be a small part of the power cost, it is a concentrated energy source, and is easily integrated into existing electric-utility systems.

Controlled fusion would provide an enormous energy supply. The cost of fuel (lithium) would be a very small part of the total expense. The scientific feasibility of this process is still uncertain, but hopefully may be established in the coming decade.

The obstacles to the commercial development of fusion power are formidable. The engineering problems which must be overcome before a commercial plant could operate are enormous. Not the least of these problems is the fact that the structure of a fusion reactor, subject to damage from high-energy neutrons, would become very radioactive and would have to be replaced every few years. With today's technology it is still not possible to assess the probable plant cost and reliability which would determine the economic viability of a fusion reactor. The availability and cost of lithium may also be a problem especially if demands for it grow for other uses such as storage batteries for electric cars. Even if all goes well, commercial operation of a fusion power plant is not expected until the next century.

The fusion plant would be similar in many respects to the now familiar nuclear-fission power station, and would be relatively easy to incorporate into present utility operations which have fission stations.

Wind

The kinetic energy of wind can be captured and converted to electricity, and the fuel is free, but the power plant (a wind machine) is not.

The major obstacle to development of wind power stems from the nature of the resource — wind is diffuse and wind speed is highly variable. Since a steady supply cannot be counted on, a means of storing electricity is needed. The capital costs of harnessing wind power and the necessary electricity-storage facilities are high. In addition, backup generation capacity is needed to provide power during prolonged periods of calm air.

Wind-power systems are best for applications which can tolerate uneven power output such as pumping water from wells, the traditional windmill application in the US. Occasional areas of wind-generated electricity may be attractive in rural areas, but large-scale adoption of wind power seems unlikely.

A 100-MWe wind facility would have the following characteristics:

Towers each 60 m high (20 stories).

Each tower has a 67-m rotor.

Twenty square kilometers of land needed (100 times that of a nuclear plant).

Average wind speed must be about 10 m/s for 100-MWe.

Towers must withstand peak wind speeds of the area (must shut down or "feather" rotor at 25 m/s).

Six hours of storage capacity needed to meet reliability requirements.

The power output of a windmill varies in accordance with the following formula as illustrated in Fig. 3:

$$\text{Power} \propto (\text{windmill diameter})^2 (\text{wind velocity})^3$$

Thus a drop in windspeed from 14 to 7 m/s reduces the power of a 60-m windmill from 1600 kW to 200 kW.

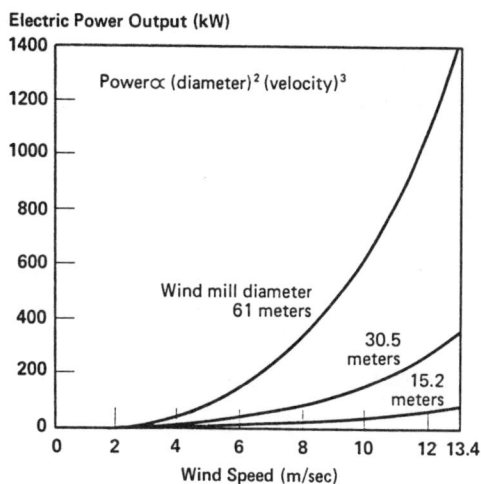

Fig. 3. Electric power generation via wind machines (effect of wind speed on electric power output).

Ocean Thermal Gradients

The difference in the temperature of water at the ocean's surface and several thousand feet deep suggests the possibility of operating a baseload power plant which would use the warm surface water as an energy source. The fuel is free since the surface water is warmed by the sun and the inherent storage capacity of the ocean is available 24 hours per day.

The efficiency of a power plant depends on the difference between the hot and cold sources utilized. For this reason, an ocean thermal power plant would be extremely inefficient because of the small temperature differential which under the best case is about 20°C. This is of no concern as it relates to the free heat, but it does require an enormous structure to obtain relatively small amounts of power.

Under current designs, warm water from the ocean surface is pumped to a heat exchanger, where a working fluid (ammonia or propane) is boiled. The vapor thus created passes through a turbine to produce power, and is then condensed in a second heat exchanger which is cooled by cold water pumped up from about a 500-m depth.

To obtain 1000-MWe of power from this type of system, probably by using ten 100-MWe modules, the warm water covering 120–250 km² of ocean would be needed. Again, the ocean is essentially free, but this serves to give some perspective of the necessary size of such a station. In addition to the requirement that each module be about 500 m deep, the heat engine itself would be enormous. One proposed design for a 120-MWe ocean thermal plant has an upper module 240 m in diameter and 30 m deep, with a cold-water-intake pipe 500 m deep and 30 m in diameter. The warm and cold ocean water would be pumped at about 6,000,000 l/s for 1000 MWe of power (this is about one-third the flow rate of the Mississippi River or about 40% of the total rate of water consumption in the US). The plant would be located 30–250 km offshore so that the necessary water depth would be available. This will necessitate a long and costly underwater transmission line.

In addition to the great expense associated with such a large plant, and the 30–250 km of underwater high-voltage transmission, the engineering task of designing a structure to operate reliably for 30 or 40 years in a corrosive marine environment is formidable. The environmental impacts of an ocean thermal power plant are not known, but several potential impacts can be identified. Mixing deep ocean water with surface water will vary the oxygen concentration of the ocean. The impact upon marine life is not known. To prevent fouling of the pumps and heat exchangers by marine growth, materials which inhibit growth will be needed. The impact of large quantities of these materials on marine life must also be considered.

Cost estimates for ocean thermal converters are very uncertain because the net power produced by such a power plant is highly sensitive to a number of design features. A small drop in the temperature difference across the power plant results in a proportional drop in gross power produced. The net power, however, is equal to gross power minus the energy needed for pumping. This pumping energy is a substantial fraction of total energy, and is a fixed load that does not change with the temperature gradient. As a result, net power will drop more rapidly than the temperature gradient, on a percentage basis. The theoretical Carnot cycle efficiency of such a heat engine with approximately 20°C of temperature difference is 6–7%. In actual operation, after allowing for pumping, the efficiency might be 2–3%. Small changes in heat-transfer rate and pumping efficiency would substantially alter plant size and cost. These sensitivities make cost estimates highly questionable at present.

In view of these obstacles and the probable high cost, ocean thermal power is not expected to provide a significant contribution to electricity generation.

Power Costs in the Future

The cost of producing electricity is an important criterion that a utility considers before undertaking the construction of a new type of plant, and the costs of producing power by each of these options will greatly influence the chance for economic success of a new option and the integration rate should these systems become economic.

The cost of electricity from these new power-generation options is largely determined by plant capital costs and capacity factors. Fuel makes at most a small impact in each case. The anticipated cost of electricity is illustrated in Fig. 4. Geothermal power from dry steam or hydrothermal reservations is the only one economically attractive today, although the breeder is expected to be competitive when it is available. It is evident that the continuous availability of power is as important as the capital cost.

Power from wind and ocean thermal gradients will be more expensive. Storage of electricity will increase the capacity factor of solar and wind power, but the capital costs will also increase, so the power costs may not be reduced.

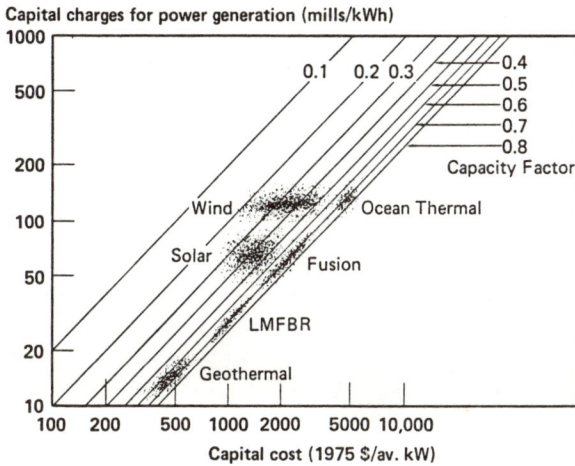

Fig. 4. Capital costs and capital charges for new electric power generation options.

Expected Paths for New Options

The development phases — scientific feasibility, engineering feasibility, commercial feasibility, utility integration, and significant use — are illustrated for several new options in Fig. 5. Historically 30–50 years has been required for new options to develop from scientific feasibility to significant commercial use, as is the case for the nuclear light-water reactor. The range of uncertainty for future developments depends upon the number of unforeseen difficulties and the intensity with which development is pursued.

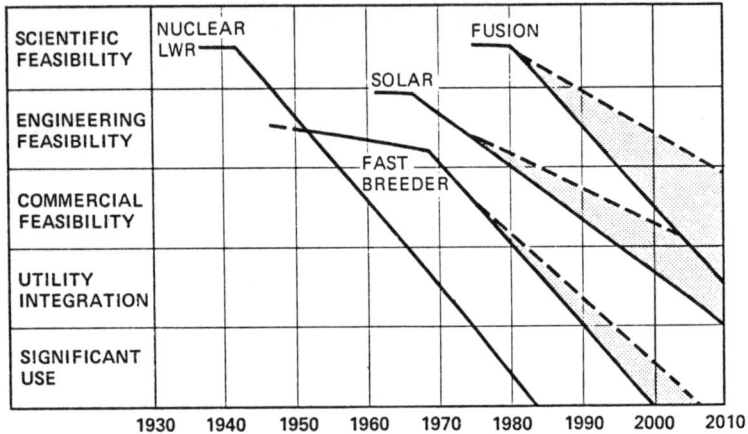

Fig. 5. Development phases for future power options.

Conclusions

The conclusions of this examination of new electric power options are:

(1) No new solutions are in sight for the next two decades.

(2) Coal and uranium will be the primary sources of electricity in the US for the balance of this century.

(3) Only minor national contributions from solar electric and geothermal can be expected by the year 2000, although they could make a modest contribution in the southwest United States.

(4) Wind and ocean thermal will not be economically attractive on a large scale.

(5) The earliest contribution of fusion might be in the 21st century.

(6) If actively pursued, the breeder reactor could be approaching significant use by the year 2000.

Certain technologies could have their rate of development accelerated by changes in national allocation of resources, by technological breakthroughs, and by changes of national policy. However, it is doubtful that the time at which these would become significantly useful can be substantially altered by such changes. It is now a cliché that money alone is not sufficient. A proliferation of pilot plants and demonstrations is no substitute for well-planned development directed at advancing the key state-of-the-art and resolving fundamental issues. Debugging a demonstration or prototype plant is more difficult than finding the solution to a fundamental problem in a laboratory.

This summary of the difficulties and long lead times that new concepts face should not be taken as an argument for abandoning advanced high-risk technologies. All such options must be explored, for history indicates that some may be successful. Because of the lead time required, a contribution to society's welfare some decades hence using advanced concepts requires that development be started now. It is obvious that the more speculative concepts will primarily benefit the next or succeeding generations. It should not be discomforting that these long-range projects do not represent technological fixes for today's problems. Some may provide such fixes for the crisis problems of the future; and it is gratifying to hope that today's efforts may provide future generations with a beneficial endowment.

6

Technical Innovation: The Answer to Resource Depletion*

Abstract

The world has always known resource limits — the predominant historical limits have been food and water. These limits have been overcome in two ways — geographic expansion and technical innovation. The possibility of geographic expansion has been largely exploited, but a wide range of technical solutions to resource pressures remains. Technical development represents the only truly unlimited resource. The concerns associated with projected resource shortfalls have always been a result of a failure to recognize the important contribution of technical development. The global problems that we now face should not be viewed from a "doomsday" perspective, but rather as challenges and opportunities for stimulating technical innovation.

Introduction

In addition to the fundamental factors of population, natural resources, and capital formation in the production activities of any given society, there is the vitally important fourth factor of technology. Here, I believe, we can find an optimistic note for the developed and emerging economies of the world. This is particularly pertinent to any consideration of the world's future energy resources.

By the year 2020 world energy demand is expected to be between three and four times present consumption if average economic growth is similar to that achieved in the past 40 to 50 years.[1] This is the result of a doubling in per capita energy consumption associated with economic growth,[2] and a doubling of world population.[3] On a global-distribution basis, nonindustrial regions (nonCommunist) now include some 50% of the world's population but use only some 15% of the annual energy. By the year 2020 their share of population is expected to be about 65%, and their share of energy could be as high as 25% if there is continued economic growth. And during the next 45 years it is expected that developed economies will produce and consume more energy than in all their previous history.[4]

*Presented at the International Conference on Energy Use Management, Tucson, Arizona, 24-28 October 1977.

Although these long-range forecasts may be considered speculative, with large uncertainties in judgement and analysis, the problems of a depleting fuel-resource base and expanding energy demand appear formidable intellectually, economically, and socially. And the tasks of seeking and implementing solutions will be enormously difficult.

What we hope to show in this paper is that the often overlooked factor of technological innovation provides a key dimension to meeting societies' expanding needs for energy. We look at this factor in a global context. Of course, whether new technological options are actually used is an economic, social, and political matter involving human values, perceived costs, and trade-offs, and will depend on the existence of social institutions structured to advantageously exploit such new technologies. In addition, we point to the responsibility of capital intensive countries to develop and advance new energy technologies.

The Growth of World Energy Demand

The growth of world population over the last two centuries has been staggering and more people are coming. During the period of Augustus Caesar there were approximately 250 million people (Table 1) growing to nearly 4 billion by 1975. And according to UN demographers, this population is expected to grow to some 6.4 billion by the year 2000, and to 12 billion by the end of the twenty-first century. Figure 1 provides a breakdown and forecast of the eight

TABLE 1
World Population, 0–2075

Year	Population (millions)
2075	12,210
2050	11,163
2025	9065
2000	6406
1975	3967
1970	3610
1960	2986
1950	2501
1900	1650
1850	1262
1800	978
1750	791
1700	602
1650	513
0	250

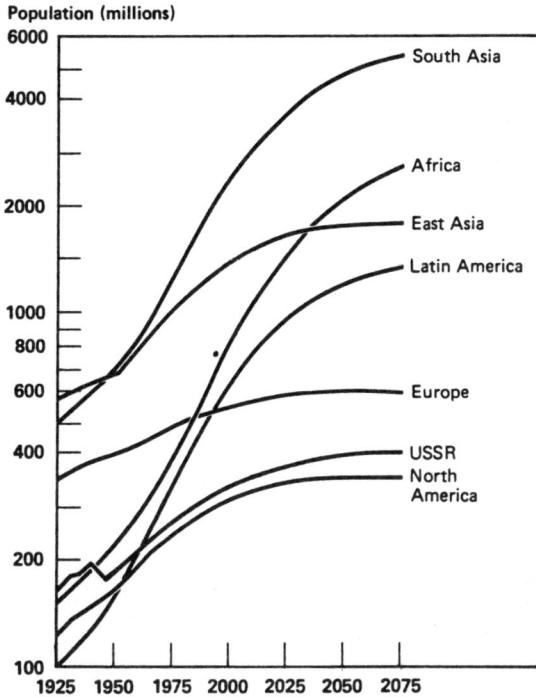

Fig. 1. Population in major world areas, 1925-2075; Medium variant of long-range projections (log scale).

major areas of the world. Confining our attention to the future only, the UN is projecting that in the century 1975 to 2075 the population of Europe will augment by one-quarter, North America and the USSR by about one-half each, East Asia by three-quarters, Latin America and South Asia fourfold, and Africa more than five-fold.[5] One needs only to review the works of Malthus[6] to understand the fate of a burgeoning population without expanding frontiers, geographic or technological. Historically, technology has provided a means of supporting the population expansion, but its implementation is a gradual process. For many emerging nations, the absence of population control has resulted in running very hard just to stand still.[7]

The economic, social and political goals have been, and will continue to be, how to feed, clothe, and house this population, and to provide opportunities for growth and development. The central element of this problem is that of scarcity, i.e. the inability of a society to provide all the food, housing, amenities, automobiles, missile systems which its members desire. The problem of scarcity is of long standing: it has kept pace with the productive contributions of science and, what is perhaps more fundamental, enormous advances in social organizations. Its persistence is due to the fundamental human trait of desiring those things that are not plentiful. Let me illustrate this with a mundane example: the feeding of people.

For more than a century, industrialized economies have appeared on the threshold of adequate nutrition, and yet, despite large advances in the per capita production of foodstuffs, the goal has not been reached — at least to the satisfaction of either the population or nutritionists. This indicates no failure of technical or economic organization, but only an expansion of man's desires. It would be easy to provide everyone with a physiologically adequate diet,[8] e.g. a man could live year after year on the following diet, at a cost of about $240 a year in today's prices:

370 pounds of wheat flour
57 cans of evaporated milk
111 pounds of cabbage
25 pounds of spinach
285 pounds of dried navy beans
and limited protein supplements

But man insists upon luxuries such as meat, and should the economic system somehow fully satisfy his desire for meat, he will no doubt insist upon shifting to another and more expensive food — lobster or caviar.

Another example, depicting continuing increases in demand over time for the same good, is provided on the dedication page of Morgan's book:[9]

"For the grandparents...who found 110 kWh/capita/day was more than adequate.

For the parents . . . who thought 150 kWh/capita/day was about right.

For the brothers and sisters . . . who have been doing well on 250 kWh/capita/day.

And for the children . . . who will probably feel like they're having to skimp at 350 kWh/capita/day."

Over time, with increasing incomes, people have continually tried to improve their lot, i.e. to improve their standard of living. Technological advances have helped to provide ever-increasing bundles of goods and services, and to alter those constraints that have prevented societies from having more. In addition, technology and innovation have introduced an ever-broadening array of products. It should be remembered that 80% of our economy is based on products that did not exist 75 years ago.[10] This trend will continue in the absence of arbitrary limits. History does not suggest otherwise. Our contention is that arbitrary limits need not be imposed if our technical resources are free to expand.

To put the growth of human energy demand in historical perspective, early man, as a hunter and gatherer of edible food, had a primary need for energy of between 3000 to 8000 calories (12.5 to 33.5 kJ) per day. This meant that the earth without agriculture could support a population of about 10 million, where the maximum consumption of food energy of early populations was probably no more than the equivalent of 4 million tons of coal (10^{17}J) annually.[11]

Early population growth was primarily determined by the restraints of food supply. This is still true globally. Food supply in turn always has been limited by man's ability to extract it from nature. This ability has been described as passing through three major phases — as shown by Deevey[12] in Fig. 2: from early man

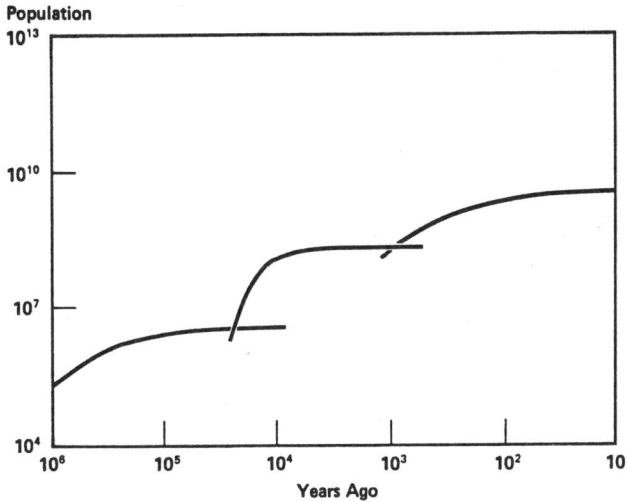

Fig. 2. World population: 1,000,000 B.C. to A.D. 1970.

whose food supply was limited by his skill as a tool user in hunting and food gathering, to his domestication of plants and animals, and to his advancement in knowledge termed the "scientific industrial revolution".[13] Each advancement in man's ability to relieve his food restraints allowed world population to take off on a new growth curve.

With the onset of agricultural technology through the domestication of plants and animals which began around 7000 B.C.,[14] total energy use increased to about 12,000 calories (50 kJ) per day, an increase of between 1.5 and 4 times more consumption than pre-agricultural societies. And by the beginning of the Christian era, the world's population of some 250 million had a worldwide energy demand equivalent to burning 150 million tons of coal (4×10^{18} J) annually. The use of energy continued to grow as a direct result of the emergence of new technologies, e.g. the water wheel and the wind mill, and the invention of the breast-strap or position harness which appreciably increased the efficiency of draft animals. And by the Middle Ages, total energy consumption was equivalent to burning about 500 million tons of coal (1.3×10^{19} J) annually, approximately 3.3 times more than in the Christian era. Although coal was employed in Europe as early as the twelfth century, it was the linking of coal to iron production, and the development of the steam engine in the eighteenth century that created enormous application of energy. By the end of the eighteenth century, the average individual in industrialized societies was consuming approximately 75,000 calories (300 kJ) per day. This corresponded to a worldwide energy demand of roughly 3.9 billion tons of coal (10^{20} J) per year, or 7.8 times greater consumption than that of the Middle Ages, and some 26 times greater than during the early Christian era, and about 1000 times that of early man. This was the beginning of the era of the Industrial Revolution.

The Industrial Revolution began in England about the time of George III, and it was a revolution by which man really began the large-scale exploitation of energy by means of machines to convert energy to useful work. It was also during this period that a major change took place in the quality of economic life, in kind and degree, because of the development and employment of modern, technologically oriented capital goods. As Ashton explains:[15]

> "Parallel changes took place in the structure of society. The number of people increased vastly, and the proportion of children and young people probably rose. The growth of new communities shifted the balance of population from the South and East to the North and Midlands; enterprising Scots headed a procession the end of which is not yet in sight; and a flood of unskilled, but vigorous, Irish poured in, not without effect on the health and ways of life of Englishmen. Men and women born and bred on the countryside came to live crowded together, earning their bread, no longer as families or groups of neighbors, but as units in the labour force of factories; work grew to be more specialized; new forms of skill were developed, and some old forms lost. Labour became more mobile, and higher standards of comfort were offered to those able and willing to move to centres of opportunity."

As shown in Fig. 3, this Industrial Revolution bids fair to conquer the globe, regardless of local race, climate, or topography, and represents a commitment on the part of societies to another way of life. Not only is there economic change associated with an industrial revolution, i.e. increases in productivity, but also cultural, political, and diplomatic change, as well as new concentrations of population, and new ways of behaving and thinking. But the key change of any industrial revolution will consist of the substitution of machines for humans and

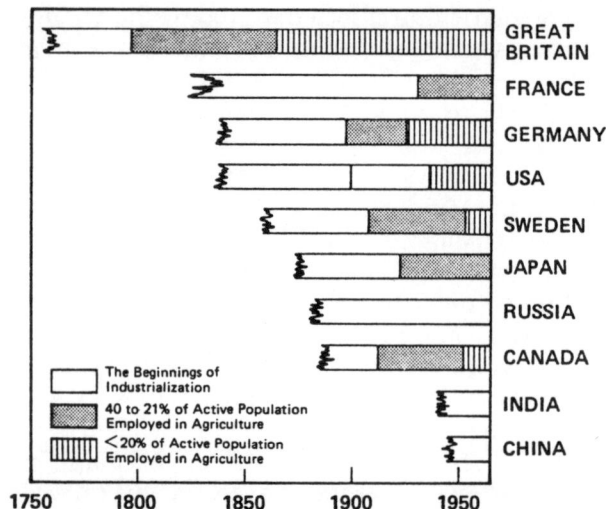

Fig. 3. The diffusion of the Industrial Revolution.

Maximum Output of Power Devices (kilowatts)

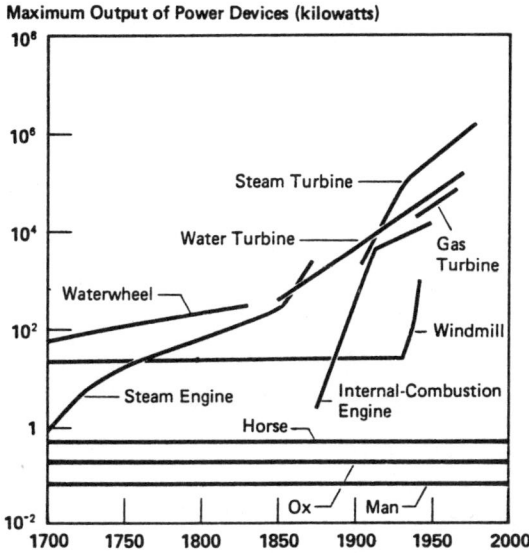

Fig. 4. Output of power devices: 1700-2000.

other work animals and the use of inanimate power on a scale never before imagined. The result is a revolution in "energy conversion". For example, the technological march from Watt's steam engine to its successor, the steam turbine, has resulted in a size expansion of more than six orders, from less than a kilowatt to more than a million (see Fig. 4). All are surpassed by the largest liquid fuel rockets (not shown) which for brief periods can deliver more than 16 million kilowatts.

It should be noted that the thermal efficiency (work output per unit of raw energy consumed) of energy systems becomes a significant issue when the conversion system costs and the fuel costs become important. The historical improvement in technology is shown in Fig. 5. The big steps were taken in the first half of the twentieth century when the economic opportunities to do so became substantial. In other words, the drive to improve the efficiency of energy resource use to perform end-use functions arises from the intrinsic costs and availability of all the components of a given system. Primary fuel cost is only one of many costs; capital cost another. History has shown us that there is nothing new about this drive for efficiency, and improvements in the future will require new materials and machine concepts that do much better.

Following the start of the Industrial Revolution, technological innovations followed each other in rapid succession. It was a revolution in technology. In this respect, the English Industrial Revolution set a universal pattern for the launching of modern economic growth and provided the key for the betterment of societies, which at all times and in all places has involved major technological innovations in the basic techniques of production. But more, technological innovation is not a random speculation on the part of scientists and engineers. It appears to be a process to finding solutions to concrete problems and bottlenecks, a challenge and response mechanism.[16]

The significance is that the past progress of technology has evolved in an essentially step-by-step fashion in a pattern dictated by the prevailing needs and conditions of a country's economy. The "kind" of technology which evolves, or is adopted, must fit both the natural resources and other factors of production. This point is of some importance when technological solutions to the world's energy needs are considered. What works best for one society may not fit the requirements of another.

In turn, the Industrial Revolution led to technological competition in which new technologies emerged and replaced old ones, e.g. the steam-powered iron ship appearing in the late eighteenth century competed and replaced the wind-powered clipper ship by the first quarter of the twentieth century. In the US,

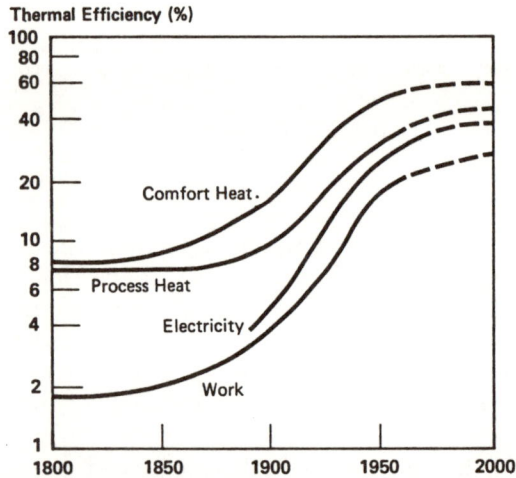

Fig. 5. Efficiency of energy converters: 1800-2000.

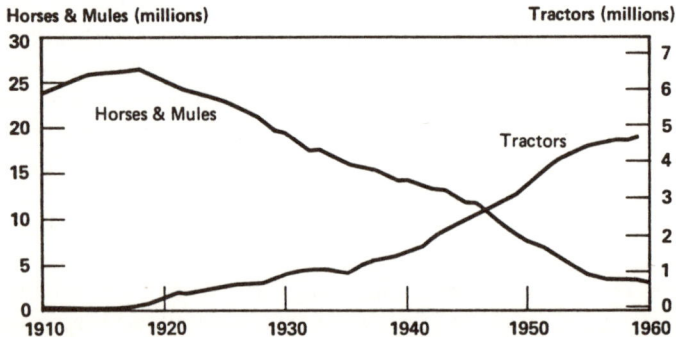

Fig. 6. Horses and mules versus tractors, US: 1910-1960.

Index (1910-1912 = 100)

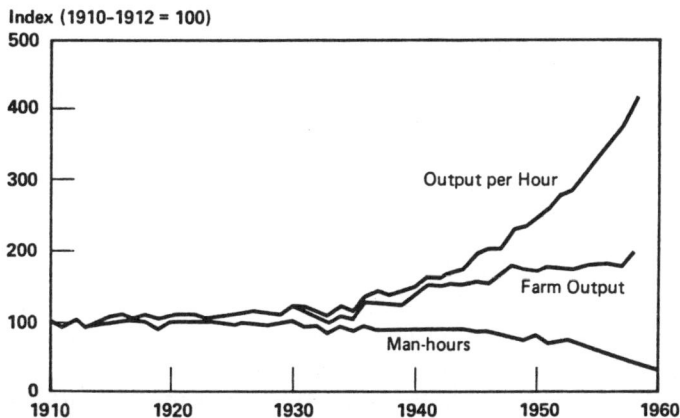

Fig. 7. Farm production, US: 1910-1956.

Energy Consumbion(kWh/day per capita)

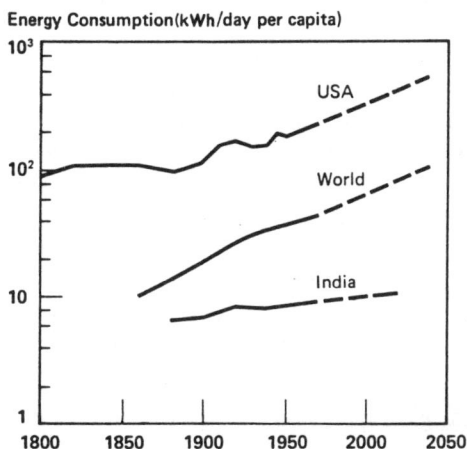

Fig. 8. Growth in energy demands: US. World, India.

machines replaced animals at a rapid rate on farms between 1920 and 1960 (see Fig. 6). Over the same time period, farm output more than doubled, and farm output per man-hour approximately quadrupled (see Fig. 7). This improvement was due not only to the internal combustion engine but also in part to higher-yielding crops, extensive irrigation, and the use of fertilizers, herbicides, and insecticides.

Today, as societies have adjusted to modern industrialization, energy expenditures continue to rise. The introduction of the automobile and airplane for personal consumption, and the development of a growing industrial base (the formation of capital) have all created enormous requirements for energy. By 1970 world energy consumption had grown to the equivalent of about 7000 million tons of coal (1.9×10^{20} J) each year, or approximately 47 times more than consumption in the Christian era. In the US the average person was consuming about 250,000 calories (840 kJ) per day, or some 21 times greater consumption than his agriculturally constrained counterpart. Those countries which have not experienced an industrial revolution have faired less well.

At present the US consumes approximately 35% of the world's energy.[17] The US share by the year 2020 will probably drop to around 25%, due chiefly to the relative population increase of the rest of the world. As shown in Fig. 8, the per capita increase in energy consumption in the US is now about 1% per year. Starting from a much lower base, average per capita energy consumption throughout the world is increasing at a rate of 1.3% per year. It is evident that it may be another century before the world average even approaches the current US level. Additionally, if the underdeveloped parts of the world could somehow reach by the year 2020 the standard of living of Americans today, worldwide levels of energy consumption would be roughly 10 times the present figure. Although these are targets unlikely to be reached 45 years hence, one must assume that developing nations are bent on moving in this direction as rapidly as political, economic, and technical factors will allow. The problems implied by this prospect are awesome.

If we look at annual resource depletion over time, it is clear that the rate of depletion has been exponential. Figure 9 provides a hypothetical example of this trend. An extrapolation from the initial data would suggest a continued exponential trend, but in the real world, we know that resources become increasingly scarce, a limiting factor that causes our curve to asymptotically approach some upper value. Technological advancements, over time, however, have continued to shift the resource-depletion curve to the right and usually forestall societies from reaching this limiting factor.

In addition, easily found and rich resources have low costs. Alternatively, as these rich resources are "used" up, low-grade resources (which may be in larger

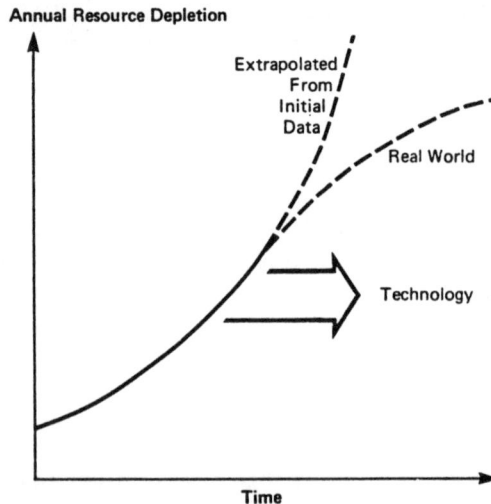

Fig. 9. Annual resource depletion extrapolated from initial data and real world.

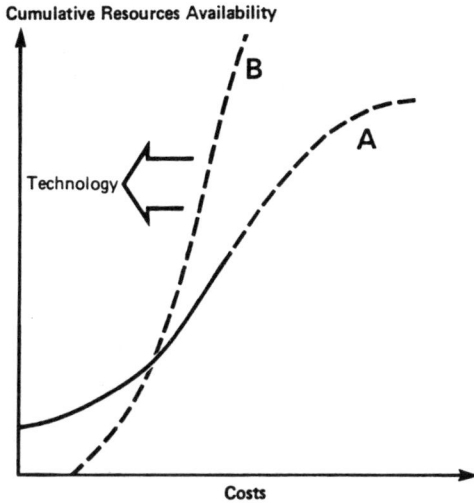

Fig. 10. Resource availability and costs with shifts due to technology.

pockets) are exploited, but only at higher costs. This is shown in Fig. 10, where curve A represents one energy form, say oil, and curve B represents another energy form, say shale-oil. In the static sense, as the cost of oil continues to rise, at some point the production of shale-oil should become attractive. But technological development is likely to shift these curves to the left, so that an alternative resource supply becomes economically exploitable sooner, and at lower cost.

Technology has been an important key in shifting these curves (Fig. 9 and Fig. 10) so that industrialized societies have not met with serious constraints, or shortages of energy. And there is no reason to believe that technological growth will stagnate, or that the system will be any less dynamic in the future. Evidence to support this thesis is provided in a later section.

A Global Energy Problem

It is important to recognize that resource depletion is a global problem.[18] Too often this fact goes unrecognized, particularly when it comes to understanding the plight of emerging nations in securing sufficient energy to develop.

The majority of the people in poor countries are at present essentially bypassed by the commercial energy market. They rely to a very large degree on human and muscle power, sunshine, and locally gathered wood or animal waste for their energy supplies. What they need, of course, is to somehow secure the supply of low-cost energy to support their development programs and to lower their increasing expenditures for imported oil and petroleum.

Increasing payments for imported oil and products is likely to increase further as developing countries industrialize and expand their energy use, and as OPEC continues to raise the price of oil. This combined with the competing demands of industrialized nations, who can easily outbid developing nations, could limit to a large degree their participation in the market. Tremendous economic pressures will exist to find substitutes, and as the tensions become significant, alternative technologies become attractive.

Many developing countries believe that their only realistic alternative to dependence on petroleum to meet conventional large-scale energy needs is the rapid development of nuclear power, and coal, and on small-scale renewable sources of energy such as wind, methane gas, sunshine, and other non-conventional energy sources. Of these, only nuclear and coal can provide energy at a foreseeable cost which will permit developing countries to compete in the international trade of industrial goods.

Much has been written about simple small-scale energy systems for under-developed areas,[19] and for rural areas these small-scale systems appear quite attractive for a number of reasons — much smaller initial investments are required, less specialized training is needed for construction and maintenance, and the cost of energy transportation is reduced or eliminated. However, while these small systems will improve rural amenities, they make little contribution to GNP. The economic incentive to seek development first in urban areas, and to concentrate on energy for production, leads to a type of energy demand that is much different than that of the rural areas.

A recent editorial in *Science*[20] which recommends rapid development and utilization of small solar systems for rural areas indicates what some of the typical costs are:

> "The cost of utility power in the United States averages 3 to 10 cents per kilowatt hour. It runs as high as 45 cents/kWh in urban areas of developing countries. In rural areas, however, power is available only from diesel generator sets at $1/kWh or more, or from primary batteries at about $12/kWh. Complete solar-thermal power systems costing about $4 per peak watt and capable of providing electricity at less than $4/kWh are already available. Photovoltaic systems costing $1 to $2 per peak watt are expected by 1980. The OTA report* says that solar devices capable of providing on-site power at much less than $1/kWh could be produced in the next few years."

The cost of urban electricity can, with sufficient load density, be significantly cheaper than "45 cents/kWh". For any rural production process, an energy cost of $1/kWh is an enormous competitive disadvantage. (And these cost estimates did not include energy-storage costs.) At this cost, a simple light bulb, run for 10 hours a day, uses over $300 of electricity a year. In many underdeveloped and developing countries, a GNP per capita of less than $300 is common. It would take all of the economic output of the typical laborer just to pay for this energy.

This must be viewed as an investment with a very low return relative to its cost. Much more worthwhile investments can be made in urban energy systems with the same amount of capital.

The economic incentive to seek development will be in urban areas first because

---

Application of Solar Technology to Today's Energy Needs, Office of Technology Assessment, Washington, DC, June 1977.

of lower costs, and second there are advantages to having centralized energy systems which meet the demands for mechanical motion (as from motors), high-temperature heat, and light. These requirements can be satisfied by electricity, but not by low-temperature energy sources such as solar space heating.

It serves to point out that the only costs that are known with certainty are those for systems that are commercially available; the costs of new technological energy sources are quite uncertain. Some options are expected to become much less expensive due to research and development, e.g. solar photovoltaic, while other options show less promise for reducing present cost estimates. Ocean thermal gradients and wind power are examples. Based upon the present level of technology, Fig. 11 provides a general cost classification of all the significant energy sources now in use or under technical development.

For developing economies, technological options to meet their energy requirements are not likely to be low cost.[21] Nevertheless, the US and other industrial countries can aid in helping developing countries to gradually transcend their dependence on imported oil, and to develop secure energy sources. Certainly, the technology to develop their under-utilized deposits of oil, natural gas, coal, and hydro-power is already available.

Low	Medium	High
DIRECT ENERGY	DIRECT ENERGY	DIRECT ENERGY
Coal	Biomass conversion	Electrolytic hydrogen
Oil	Coal liquefaction	
Gas	Wind for pumping	ELECTRIC CONVERSION
Farm waste		Wind electric
Solar heat	ELECTRIC CONVERSION	Ocean thermal
Coal gasification	Solar-thermal electric	Satellite solar
	Fusion	Photovoltaic solar
ELECTRIC CONVERSION	Diesel	
Coal	Biomass	
Fission converter	Oil	
Fission breeder	Gas	
Hydroelectric		
Geothermal		
Municipal waste		

Fig. 11. Relative cost of energy production (excluding transportation).

The Growth of Technology

As pointed out earlier, primitive food gathering societies were limited by the natural rate of food production, but domestic agriculture removed this limitation. Agrarian societies were constrained by the limits of land, but the Industrial Revolution provided other opportunities. In the US 100 years ago a fourth of all arable land was used for feed for draft animals. At that time, an extrapolation of agricultural trends a century ahead would have been extremely pessimistic. Consider this quote from 1922.[22]

"According to Eugene Davenport, Dean of the Illinois College of Agriculture, the greatest future need of American agriculture is a fundamental national policy. One of our leading statisticians estimates that a century from now our population will amount to more than 225,000,000 people. The prophecy is startling because it suggests possible hunger and even famine as our future. At present, with less than half these numbers, our food production is only about equal to our domestic consumption. Unless we institute very revolutionary practices to enhance production, we may look for it to fall behind. Present indications are that we are rapidly slipping into the class of food-importing nations. This means that unless we are able to reverse the tide, we must readjust our social, economic and industrial organizations to accord with this new condition."

The outcome did not fulfill the prophecy — the US reached the stated population in roughly 50 years, not 100, yet is the leading exporter of food in the world. Solutions were found in geographic expansion and technological developments. Today, expansion to new lands has been largely exploited (unless we consider the advocates of space colonization seriously), but all evidence indicates that the technology frontier has no limits.

Economic theorists have examined the technological component of economic growth.[23] The standard economic tool for such an examination is the Cobb–Douglas production function which relates economic output to capital, labor, and time (to incorporate technological change) and other factors. In its general form this production function has been represented as:

$$Y = e^{\lambda t} K^{\alpha} L^{\beta} ,$$

in which Y is economic output, K is capital, and L is labor. λ, α, and β are constants. In this general form, assuming other things remain the same, output grows exponentially in time. This is representative of technological change. Based on this production function it is possible to maintain growth of output concurrent with reduction in resource consumption. This appears possible because under this representation of the production process, substitution of capital and technology for other resources is possible. However, this is only valid under certain conditions, and the extent to which these factors can be substituted for one another has not been sufficiently explored.

One controversial study,[24] employing American data for the period 1909–1949, showed that the upward shift in the production function was at a rate of about 1% for the first half of the period and 2% for the last half. In addition, Solow concluded that gross output per man hour doubled over the interval, with seven-eighths of the increase attributable to technical change and the remaining one-eighth to increased use of capital.

To respond to the energy problems of tomorrow for both developed and underdeveloped economies, technological growth may be expected to grow exponentially.[25] It will require acceptance and support on the part of individual societies.

Without defining one parameter, or a group of parameters, as a measure of technological progress, it is possible to generalize from the evidence found in

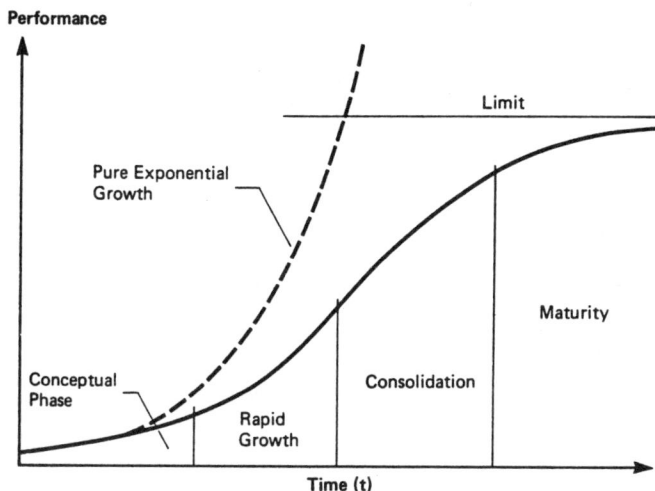

Fig. 12. General form of the sigmoid curve; Performance= Limit/$(1 + be^{-at})$.

common growth from the various technical fields that technology exhibits an exponential growth pattern. Although the techniques of technological forecasting can be quite complex,[26] there are at least two predominant characteristics one can build forecasts on. First, when the technical parameter for a specific technique is plotted against time it follows a characteristic sigmoid curve of the form shown in Fig. 12. Initially, the technique tends to experience a period of slow growth in its initial or conceptual phase, a subsequent period of exponential growth, and then a flattening as limiting factors are encountered, and the curve asymptotically approaches some upper value, which is definable when the limiting factor is known.

This initial phase of growth along a sigmoid curve can be expressed by the following infinite series:

$$\frac{\text{Output}}{\text{Limit}} = \frac{\varepsilon e^{at}}{1 + \varepsilon e^{at}} \sim \varepsilon e^{at} - (\varepsilon e^{at})^2 + (\varepsilon e^{at})^3 - \dots$$

where $\varepsilon\, e^{at} < 1$.

Because initial data in any gross situation supports the first term, the common omission is to neglect the higher terms and to extrapolate the first term until the "Doomsday" syndrome occurs.

Second, there is the major characteristic that the overall growth of a technological field is composed of a series of sigmoid curves where each specific technology contributes a small portion to the overall capability growth. This point is illustrated in Fig. 4 and in Fig. 13 to Fig. 16 where the successive generations of technological steps have increased the performance of the power output of basic machines, lighting efficiency, the rate of increase of operating energy in particle accelerators, and the performance levels of various generations of computers as a function of time. The overall exponential growth rate can clearly be seen from these figures.

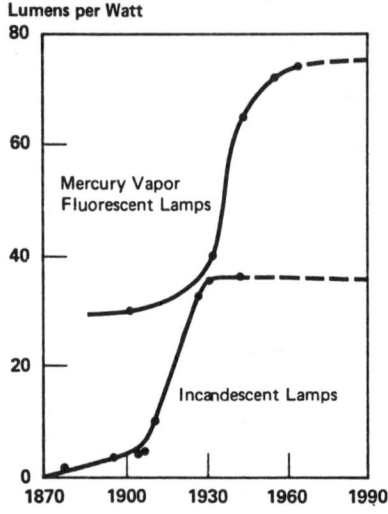

Fig. 13. Illumination.

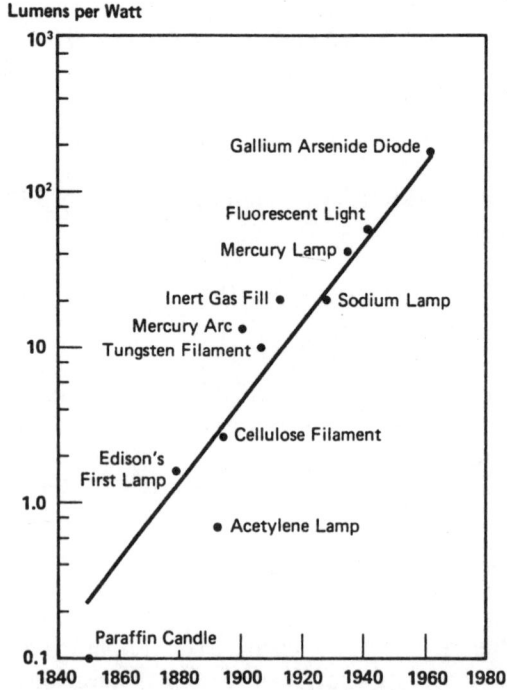

Fig. 14. Illumination trends.

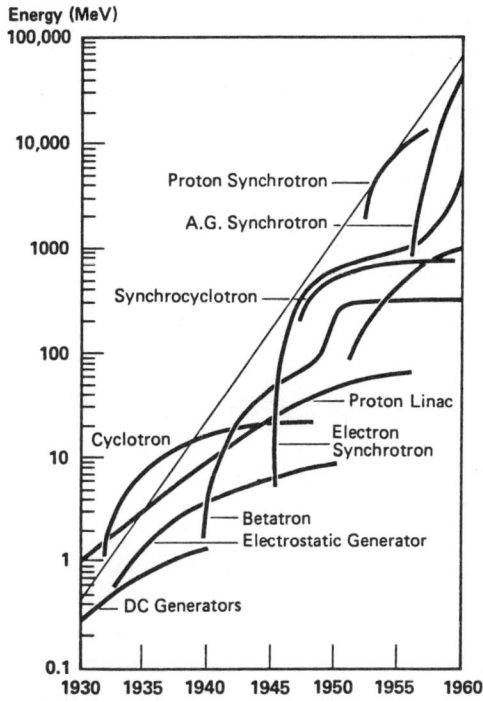

Fig. 15. Particle accelerators.

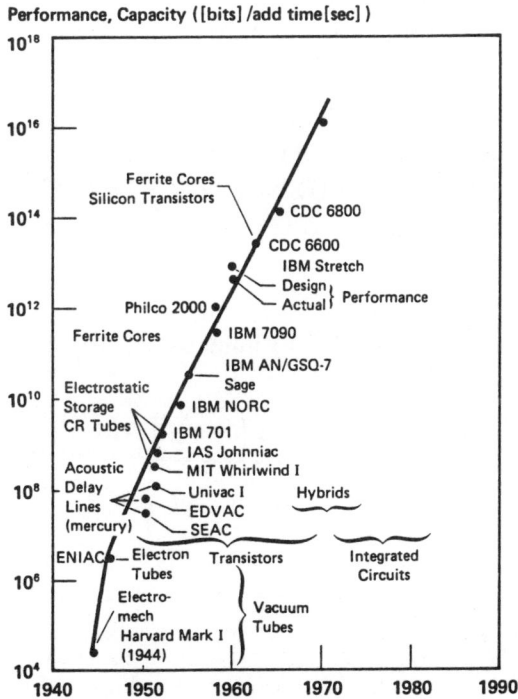

Fig. 16. Computer generations.

Although these figures are only representative of a much larger base, one needs only to turn to the technical publications and reports to have a glimpse at recent and pending technological advancements in the field of energy utilization. For example, a recent report of a NATO Science Committee Conference[27] highlighted energy problems and technological opportunities in heat, light, motion, electrolytic and electronic processes, residential and commercial systems, industrial systems, and urban systems to provide energy for a growing society efficiently. One can be optimistic that the anticipated technological advancements will serve to significantly reduce future energy requirements and also increase energy-supply alternatives.

Mathematically,[28]

"Assume that the discretionary resources of society at any real time, $t = r$, is the proportional product of a past technology at $t = p$ where r is real time and p is past time. Thus, if T is the strength of technology, then societal resources (at $t = r$) $= \alpha\, T_p$ where α is a proportionality constant. Societal expectations from investments in technology are assumed to determine the fraction of the available discretionary resources that will be allocated to technological development. These expectations are defined by two parameters. The first, the technological "payoff factor", is expressed as the relative difference between the technology (T_p) in common use and the perceived potential level of technology capability (T_r) based on current resource strength as previously defined. The second parameter is the relative priority which society assigns to attaining the performance objective associated with the perceived potential of the most updated level of technology. Expressing this mathematically, societal expectations from technology are equal to $\beta\,(T_r - T_p)/T_p$ where β is the relative priority factor and $(T_r - T_p)/T_p$ is the technological payoff factor. The amount of discretionary resources allocated can now be defined as the product of the total resources available and the societal expectations; that is, resource allocation to technology is equal to

$$(\alpha T_p)\left(\beta\,\frac{T_r\text{-}T_p}{T_p}\right).$$

If, then, the rate of change of technology is proportional to the societal investment,

$$\frac{\mathrm{d}T_r}{\mathrm{d}t} = \alpha\beta\,(T_r - T_p) = \alpha\beta\left(1 - \frac{T_p}{T_r}\right)T_r$$

$$= \alpha\beta\,(1 - \gamma)\,T_r$$

where $\gamma = T_p/T_r$. It is evident that this is a growth relationship and — assuming α, β, γ are constants — is of the simple mathematical form

$$T_r = T_0 e^{\emptyset t}$$

where $\emptyset = \alpha\beta\,(1 - \gamma)$ and thus technology exhibits an exponential growth.

If $\varnothing$ is indeed a constant, then γ can be expressed as

$$\gamma = T_p/T_r = e^{-\varnothing\,(r-p)}$$

where $(r - p)$ is the delay time for technology applied to societal needs.

"In the convenient nomenclature of doubling time, D, we have $\varnothing D = 0.693$. Conventional wisdom suggests that the technological component of economic growth has a doubling time of 20 to 30 years. For 30 years, $\varnothing = 0.023$; that is, a 2.3 percent annual compounded growth. If the delay time for practical innovation is roughly 10 years, the $\gamma = 0.79$; and thus 21 percent of potentially available technology is not yet in use. These appear to be reasonable numbers for illustrative purposes, although the use of a single equation for all of technology is an obvious oversimplification."

Additionally, the exponential rise shown in Fig. 17 and Fig. 18 for the number of scientists and engineers per million population and for research and development funds as a percentage of the GNP raises an analytical question not addressed in the mathematical analysis.[29]

"It was there assumed that the fraction of societal resources applied to technology was primarily determined by the technological delay time rather than by an increasing importance of technology in the social system, that is, for scientists and engineers per million population

$$\frac{R\ \&\ D}{GNP} = \beta\left(\frac{1}{\gamma} - 1\right)$$

Number of Scientists & Engineers per Million

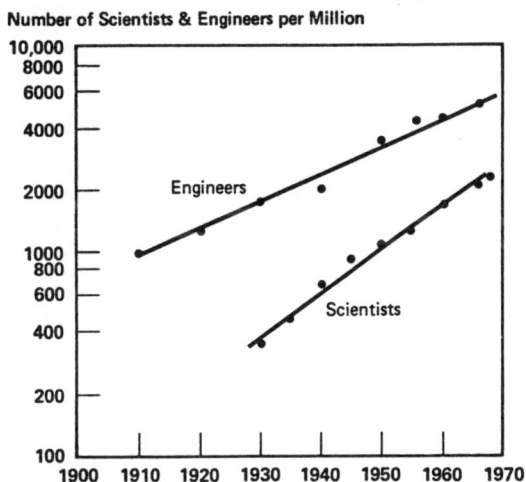

Fig. 17. Growth trend for scientific and engineering personnel in the US.

Total R&D Fund as Percent GNP

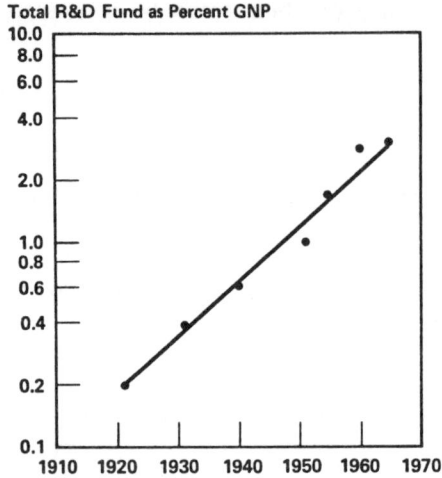

Fig. 18. Trend of research and development expenditures as a percentage of GNP.

"Thus our elementary mathematical approach produced an exponential growth in technology even with a fixed fractional allocation of available resources. Inherent in this approach was the assumption that the new technological options would increase the total resources available and thus result in a continuous growth of technology. The role played by the increasing percentage of the GNP allocated to research and development during the last half-century was not considered in the analysis and would represent an additional forcing device for technological growth. This question should be examined further. However, the key point is that even with a fixed percentage of the GNP directed to new technology, an exponential growth should occur."

But could it not be expected that technology as a whole, over the long run, be approaching an inevitable inflection point?[30] I think not because we are so far from the theoretical limits of performance. Of course, it is impossible to ascertain which portion of this aggregate technology curve we are on, since technology development should be a sigmoid process. Intuitively, I believe, we are in the growth phase, and if there are limits they lie far in the future. Perhaps those limits will be reached when societies reach a state of omniscience, or when societies lose interest in the reinvestment process, or become non-technical — like the porpoise, intelligent but without machines. These states, however, seem quite remote. Perhaps a more important question is how technology and societies interact. We now turn our attention to the relationship between the "well-being" of societies and technology over time.

Progress of Technological Societies

Unfortunately, the art of societal assessment of the impact of technology is at best rudimentary simply because those factors which would be most indicative of technological and societal status today or in the past are missing. However, one attempt by Gordon and Shef[31] does provide some insights into the growth of technology over time, and social and economic growth as a function of technological progress.

Developing a composite index for technology composed of power, steel, transportation, engineers and scientists, and communications, it was shown that the technological status of the world as a whole advances at a roughly constant exponential rate (see Fig. 19), doubling every 20 years. Although slight temporal differences exist, the technological growth rate from at least the beginning of the twentieth century has been relatively constant for the world as a whole. The technological status of the world today is roughly equivalent to the US at the beginning of the twentieth century. California is included as a subnational society of the US and serves as an example of the extreme rapidity of growth of technological status which can occur in subsets of individual nations. Japan serves as an example of a country that has crossed over the boundary from the underdeveloped technological group to the advanced group. Worthy of some note is China. If this index reflects true technological progress, then China is probably experiencing the most rapid expansion of any country within this century.

Relating a social-economic composite index (composed of such values as GNP per capita, university students, infant mortality, mental health, unemployment, urban crime, and so on) to the technological index, it appears that there is a one-to-one correspondence between technical progress and social/economic progress in the middle regions of Fig. 20. One, of course, has to be extremely careful in not pushing the correlation of this figure too far, but it does provide a striking example of a "stream tube" through which countries have had to pass on their way to a post-industrial world.

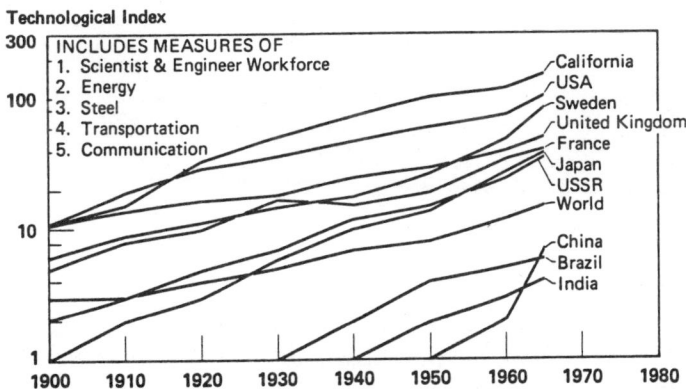

Fig. 19. Technological index as a function of time.

As shown in Fig. 21, the pattern of approach in the technological-social/economic domain appears to vary over time. Pre-industrial countries early in the century approached on path A, whereas path B appears typical of the progress of emerging nations in the past few decades. Path B suggests that these countries have been able to borrow and adapt technological capital.

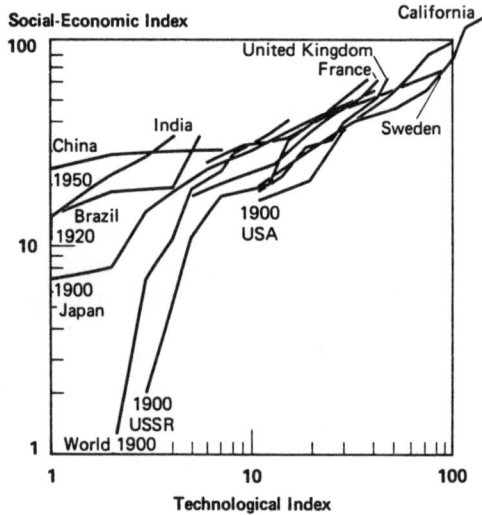

Fig. 20. Social and economic growth as a function of technological progress.

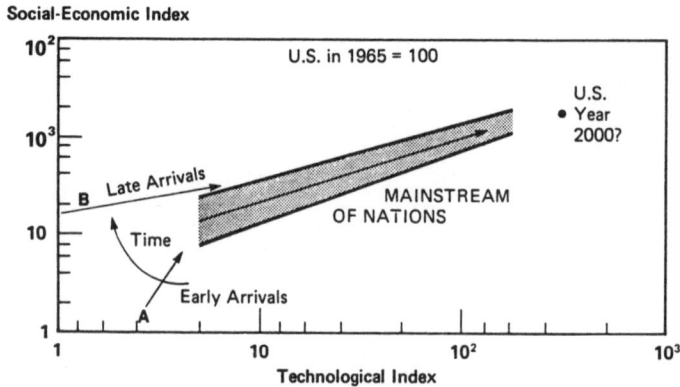

Fig. 21. Patterns of social/economic and technological progress in the twentieth century

Conclusions

There has been a great deal of lamenting associated with the projected resource shortfalls. One could even argue that it has become a national pastime. But there is hope, particularly for developed economies, if attention can be focused on the important factor of technology. Of all the resources available, it is the only one that is man-made. And armed with the stock of information accumulated over the centuries, technological solutions to resource-depletion problems will be found. To be sure, it is an enormous challenge but there is no reason to believe we cannot meet it.

Our life styles and hope for continued improvements in the quality of life need not necessarily be abandoned, or changed. There is no need to go on a strict diet of fewer calories unless, of course, the members of this society want to, and it does not look like they want to. There is that choice available to us. And there is every indication of being able to open up new technological frontiers for continued social development in both the industrialized and developing nations of the world. Let us focus our attention on this unlimited resource and create the technologies to meet the world's future energy requirements.

References

1. I. Bloodworth et al., "World Energy Demand to 2020", A discussion paper prepared for the Conservation Commission of the World Energy Conference by the Energy Research Group, Cavendish Laboratory, University of Cambridge (15 Aug. 1977).
2. C. Starr, Energy and power, Scientific American (Sept. 1971).
3. United Nations, Concise Report on the World Population Situation in 1970-1975 and Its Long-Range Implications, U.N. Publications, New York, 1974.
4. I. Bloodworth, op. cit.
5. United Nations, op. cit.
6. T. Malthus, J. Huxley and F. Osborn, On Population, The New American Library, New York, 1960.
7. R. Gill, Economic Development: Past and Present, Prentice-Hall, New Jersey, 1967.
8. G. Stigler, The cost of substance, Journal of Farm Economics, XXVII (1945).
9. M. Morgan, ed., Energy and Man: Technical and Social Aspects of Energy, IEEE Press, New York, 1975.
10. J. Lasser, How to Run a Small Business, McGraw-Hill, New York, 1974.
11. H. Brown, Energy in our future, Annual Review of Energy, 1 (1976).
12. E. Deevey, The human population, Scientific American (Sept. 1960).
13. Ibid.
14. C. Cipolla, The Economic History of World Population, Penguin Books, England, 1974.
15. T. Ashton, The Industrial Revolution 1760–1830, Oxford University Press, London, 1962.
16. W. Rostow, The Process of Economic Growth, W. W. Norton, New York, 1962.
17. C. Starr, Energy and power, op cit.
18. J. Sewell, The United States and World Development, Praeger, New York, 1977.
19. A. Makhijani, Energy Policy for the Rural Third World, International Institute for Environment and Development, London, 1976.
20. A. Hammond, An international partnership for solar power, Science, 197 (1977).
21. C. Starr, Energy systems options, paper presented at the 10th World Energy Conference, Istanbul, Turkey, 1977.
22. Scientific American (Dec. 1922).
23. R. Allen, Macro-Economic Theory, St. Martin's Press, New York, 1968.

24. R. Solow, Technical change and the aggregate production function, *The Review of Economics and Statistics* (Aug. 1957).
25. C. Starr and R. Rudman, Parameters of technological growth, *Science*, 182 (1973).
26. J. Bright and M. Schoeman, eds., *A Guide to Practical Technological Forecasting*, Prentice-Hall, New Jersey, 1973.
27. E. Kovach, ed., *Technology of Efficient Energy Utilization*, The report of a NATO Science Committee Conference (1973).
28. C. Starr and R. Rudman, *op. cit.*
29. *Ibid.*
30. D. Price, *Little Science, Big Science*, Columbia University Press, New York, 1963.
31. T. Gordon and A. Shef, National programs and the progress of technological societies, presented to the American Astronautical Society (March 1968).

ENERGY ANALYSIS
AND PLANNING

7

*Energy and Society**

Introduction

It is now common wisdom that energy is a crucial need of our society. International events of the past few years, by causing sharp perturbations in our flow of resources, have illuminated our dependence on energy. Unfortunately, public discussion of this subject has been clouded by political overstatement and analytical understatement.

Analysis of the historical influence of energy availability and cost on the development of our industrial society indicates that the effects are substantial and diverse, and that there are probably no very simple generalizations that can provide a basis for national policies. The variety of energy resources, forms, and end uses is so great that there exists a broad spectrum with regard to societal needs, efficiency of use, importance to public welfare, and flexibility to change. As with national food consumption patterns, we have the lean and the fat, good nutrition and malnutrition, the efficient and the wasteful, the cheap and the costly. No single formula can apply to all these cases. What is clear from energy-system studies is that our quality of life, material welfare, health, employment, and income are demonstrably affected by energy availability and cost; and we now know that there will be no return to the cheap abundant energy of the past. These developing constraints on energy use may be key causative factors in increasing unemployment, reducing real income, and otherwise deteriorating the material aspects of our society.

What is at issue is our ability to rationally control these matters so that we achieve a good national mix of outcomes. In current political terms, is it feasible to adjust our energy systems so that we simultaneously decrease unemployment, decrease costs, decrease dependence on foreign resources, decrease undesirable effluents and impacts on the biosphere, i.e. land, air, and water? This paper will not provide the answer — but may shed light on some of the factors involved.

In a developed industrial society there exists a set of interrelationships that tie together energy use and a wide range of physical, economic, and social aspects of life. This is no different, in concept, from the web of interrelationships which form the ecology of natural systems. And as in these natural systems, the ecology of man-made systems can be altered in complex ways if the system is tampered with at any point. Thus when we undertake to alter consciously the availability of

---

*Presented at the Hoots Lecture Series, School of Earth Sciences, Stanford University, Stanford California, 26 January 1976.

energy or our societal patterns of energy consumption, it would indeed be desirable if we could foresee the principal repercussions of such actions on our total social structure.

Unfortunately, the art of societal assessment of the impact of proposed technologic or economic actions is at a rudimentary level in the best of cases, and is particularly embryonic in the field of energy. Although always recognized as a need, energy has customarily been considered a low-cost, widely available resource and its societal interactions have been rarely studied. It has certainly received much less consideration than food or water supply, for example, even though energy availability is almost as influential in an industrial society as the availability of these more obvious life-supporting needs. The disparity in perceived roles is evidenced by our politically accepted government subsidy for agriculture, which is now about one-third of the total US farm income — with nothing comparable for energy. It is therefore constructive, in these days of public concern with energy issues, to consolidate some of the insights that the studies of this subject have developed. Hopefully, these may suggest some important causal relationships that will be of value in assessing new energy policies.

The major societal effects of energy systems developed by man can be related to four predominant technical sectors: first, the use of work animals; second, the direct conversion into mechanical work of inanimate stored solar energy (hydropower, wood, farm waste, and fossil fuels); third, the use of electricity as an energy form derived from these sources; and fourth, the use of nuclear energy as a primary source. The past and potential impacts of these developments are so many that only a few principal aspects can be explored here: chiefly those relationships of energy flow to the societal materialistic targets of health, security, social stability, upward mobility, and leisure.

Although a more complete list of societal targets would undoubtedly include spiritual as well as these materialistic goals, the above are most directly affected by energy relationships. The traditional target of "economic growth" may be one of the best composite means of achieving all of these materialistic goals, and this explains its frequent and popular use as a measure of societal success. Thus, industrial growth has historically been a prime objective. Other contributors to social development have been greater productivity in agriculture, improved transportation and communications, and shorter working hours.

I. The Industrialization of Society

The use of inanimate stored energy devices as a means to supplement human labor is a development only a few hundred years old. While the history of the water wheel goes back much longer, its early use was limited to assisting traditional functions such as milling grain or lifting water. Its use to drive machinery began to grow about 300 years ago, as shown in Fig. 1, when the large wheels were built and initiated the industrialization of society. The location of water streams limited the siting of such industrial activity. This limitation was eased about 200 years ago by the commercialization of the heat engine which permitted portable fuels to be used to drive machinery.

Maximum Output of Power Devices (kilowatts)

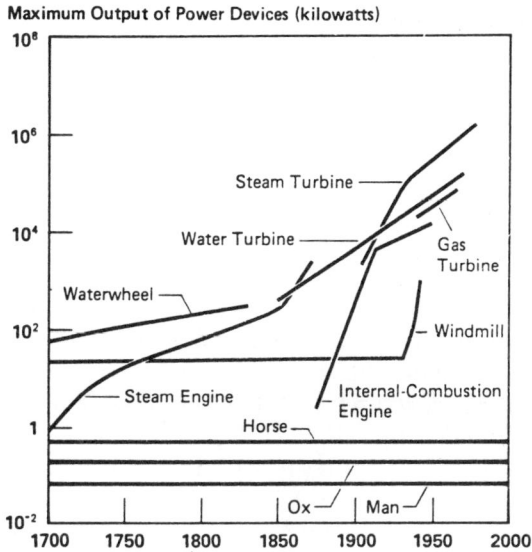

Fig. 1. Power output of basic machines.

Because of the physical bulk of the early heat engine, it was most convenient to locate industrial machinery in groups driven by single engines. Thus, the typical factory a century ago was a large structure containing a steam-driven engine which turned a single main shaft extending the length of the building. Leather belts transferred the power from the main shaft to individual machines. This resulted in a very low efficiency of energy use, in part because the steam engine and main drive operated continuously, even though the individual machine operations were intermittent. In addition, the belt-drive system required the work to be done near the main shaft, rather than in a more efficient assembly area. Thus, both efficiency and mobility were limited.

The development of the electrical generator and motor in the late 1800s provided a radically new flexibility to the distribution of energy for mechanical work purposes. Free of the need for mechanical linkage, work could be done where and when needed. The result was a substantial increase in the efficiency of industrial work output per unit of energy consumed.[1] In a similar fashion, the development in the early 1900s of the internal combustion engine (diesel and gasoline) provided a similar flexibility and resultant expansion of use in transportation systems and mobile machinery.

The effect of these developments on the structure of industrial societies is much more significant than would be indicated by the economics of energy supply. In the US, the cost of primary energy fuels (crude oil and gas at the well, coal at the mine, etc.) is only a few percent of the national product; and the cost of energy in all forms delivered to the end-using device (gasoline at the pump, electricity at the user's switch, gas at the home, etc.) represents only about 10%. These modest relative costs do not properly indicate the leverage of energy systems on the pattern of societal development.

It is instructive to illuminate the influence of the above changes in available energy options by examining the transitions in energy use during the past century. These have been reviewed in detail for the US in reference 1. The most general transition, shown in Fig. 2, is the worldwide shift in the fuels mix which began a century ago. In the US, fuel wood shifted from half of all fuel to less than a percent now (Fig. 3). The compensating increase in the use of mineral fuels provided very concentrated abundant energy sources, and thus facilitated the unrestrained expansion of energy uses which would not have been even feasible if wood had been the only resource.

The US history of fuel wood illustrates several characteristics of energy systems. Wood, waterpower, and windpower were the principal energy sources two centuries ago. Wood was the predominant source, with domestic heating consuming over half. Its cost was very low, because it was a superabundant by-product of the then extensive conversion of virgin forests to agriculture.

The abundance of wood in the first half of the last century, and the generally rural distribution of the US population, resulted in the economic situation of a practically limitless fuel supply (compared to demand) at very low cost. The result was extravagant and inefficient use for domestic purposes such as heating, smoking meat, and drying tobacco. The bulk was consumed in the open fireplace so that[2] "a poor man, even a plantation slave, could burn bigger fires than most noblemen in Europe . . . . If the fire was too hot, he left the doors open, but fire he would have if only to brighten up the dark end of the house".

The heating efficiency (heat into the room compared to the energy content of the fuel) of such an open fireplace is about 5 to 10%, depending on configuration. In Britain, at the same time, the lack of wood resulted in the use of the coal fireplace with a space-heating efficiency of about 15% (twice that of the US open wood fireplace). In Russia, the cost of fuel was such that space-heating stoves were designed for maximum efficiency; greater than 30% was achieved. The economics of wood in the US made it not worthwhile to pay for an efficient stove, or to use valuable man-hours to chop the wood into small enough pieces to fit into such a stove. To place this extravagant use of wood in perspective, it is revealing that the US per capita use of energy for household purposes of all kinds is about the same now as in 1880 — this in spite of the many household appliances and temperature-conditioning devices we now use. The reason, of course, is improved energy-conversion efficiency.

The wood fuel dominance began to disappear simultaneously with its reduced availability as compared to increasing energy demand. Wood was also bulky to transport, compared to coal or oil. It is interesting that each of the fossil fuels — coal, oil, gas — faced substantial obstacles before public acceptance (not unlike the nuclear-power situation of today). It took decades to convince wood users that coal could be used instead, both in the home and industry. It does burn in a more somnolent fashion and requires special grates and handling.

It was not until about 1900 that coal was fully accepted in the US — in contrast with Britain, where wood was scarce, and coal became a dominant fuel in about 1860. Wood has, of course, almost completely disappeared from the US scene.

Oil entered the world scene primarily as a source of kerosine for illuminating purposes. It became competitive only after whale oil, lard and tallow became

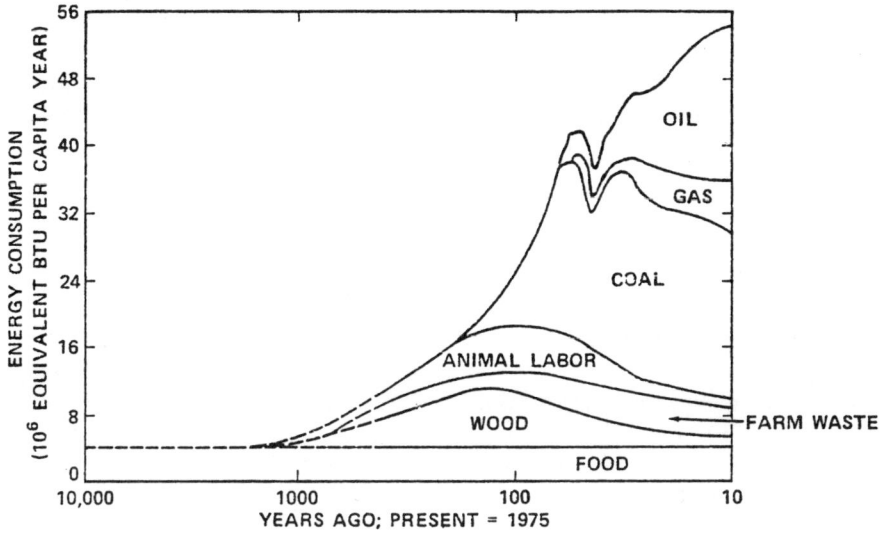

Fig. 2. Per capita consumption of energy resources in the world.

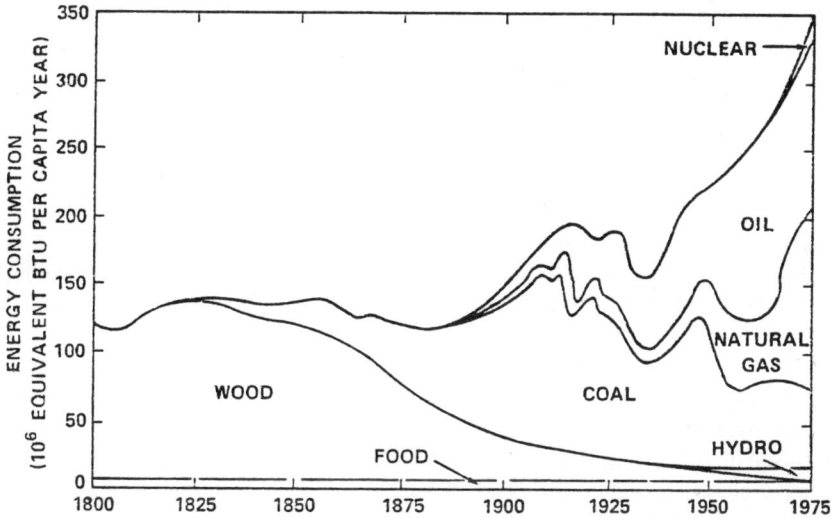

Fig. 3. Energy consumption in the US, 1800-1975.

expensive. Its expanding availability and dropping cost resulted in a general expansion of night-time illumination. Although in urban areas coal gas was generally available in the middle 1800s, rural America had no such amenity. Illuminating oil, therefore, became a major stimulus in the second half of the century to the public education movement in the US. The light level provided by the per capita consumption of illuminating oil was certainly adequate — equivalent to 8.4 candlepower hours per person per day. This had the direct effect of increasing the number of hours in the day when reading was possible and also the time available for educational activities, enough to contribute to a sharp increase in the literacy level during this period.

It was not, however, until late in the century that oil as a fuel for power generation began to play a role. In 1867 the then Secretary of the Navy pessimistically stated, "It appears that the use of petroleum as fuel for steamers is hopeless". Safety and health were big issues. By about 1900 oil as fuel for railroads, industry, and merchant vessels began to grow. Even the Navy capitulated.

The internal combustion engine was, by hindsight, the great potential user of oil. It had been around for 200 years as an experimental device, but the absence of a suitable fuel had inhibited commercial use. Gasoline made the difference. At the beginning of this century an active rivalry existed between the steam, electric, and internal combustion engines; but the convenience of the gasoline-driven car made it the winner. The diesel engine for heavy machine drives also became available. The mobile machine age was underway and blossomed in the first half of this century.

The societal impact of these technical developments is familiar to all of us: the industrialization of agriculture, the dispersion of field manufacturing operations, the expansion of commercial transportation systems — truck and air — and finally the fantastic individual mobility provided by the automobile.

The societal consequences of the personal mobility provided by the automobile have probably been as significant as the availability of power machines in industry. It has altered the demographic distribution of communities, industry, and supporting facilities such as medical facilities, schools, and retail distribution centers. It has equalized and homogenized rural and urban opportunities and social patterns. It has also created substantial energy consumption for personal transportation, which on a per capita basis, *had never existed before*. One can hardly describe the automobile as having been embedded in an existing societal system. Rather, during the past half-century, we have restructured society about the automobile's performance characteristics. At this stage of dependence on personal mobility, only alternative vehicles, such as the electric car, could possibly replace the gasoline-fueled car.

The development of the last fossil fuel, natural gas, did not create major new options. Natural gas has become a convenience fuel for heat production at high-density fixed sites — factories, residential, and commercial areas. As an energy system, it supplements and replaces coal and oil. Its greater convenience in use and environmentally clean combustion are advantages; its lesser safety and lack of packaged mobility are disadvantages. Historically, the unique illumination produced by the combination of gas and the rare-earth mantle had a short-lived use due to the advent of the electric lamp. So natural gas did not create unusual societal change, although once installed as a fuel system it is now an important resource.

The advent of electricity generators and motors opened a whole new range of system options. Electricity is an energy form — not a source. It can be generated from any of the raw energy sources, transmitted and distributed with relative ease, and converted into useful work, heat, or light at any point. Although it can be stored in small quantities by electrochemical reconversions (the storage battery), it cannot as yet be stored as electricity in quantity. This is its primary disadvantage. Otherwise, it is almost the perfect energy intermediary.

The successful development of large size rotating electrical machinery in the late 1800s started a commercial production and distribution industry that began to grow rapidly after 1900 (more than doubling every decade, i.e. in a half century almost 20 times greater growth than total energy use). Its convenience in distribution and use, as compared to any other energy form, has made it the user's first choice. Electric power stations can be built for any fuel and thus provide long-range resource flexibility without changing the installed distribution and end use devices. Its continued growth as the preferred intermediate energy distribution form is inevitable.

Perhaps one of the most important advantages of electricity is that it has the greatest potential for permitting efficient conversion of energy sources into end use functions. Electricity power plants, by their very size, permit the most efficient machinery to be used for generation. Transmission and distribution losses can be kept as low as is economically worthwhile; and the reconversion of electricity to work can be done more efficiently than with any other energy system. The net result is that for any work purpose, the electric system utilizes fuel most effectively. Only in heating does the direct use of fossil fuels compete, and even here, because of its ease of control and flexible location, electricity often does better.

It has been pointed out by Schurr[1] that the introduction of the electric motor drive into manufacturing caused the energy input per unit of GNP to drop significantly. The prime mover-belt-drive factory was one-half to one-third as efficient as the prime mover-electric-motor factory. In 1900 only 5% of industry used electric motors, and now it is more than 90%. This objective of decreasing the energy input for end-use function is, of course, a major objective of all research and development in the electrical industry.

The thermal efficiency (work output per unit of raw energy consumed) of energy systems becomes a more significant issue when the conversion system costs and the fuel costs become important. The historical improvement in technology is shown in Fig. 4. The big steps were taken in the first half of this century when the economic pressures to do so became substantial. Improvement in the future will be more difficult — primarily because the metallurgical properties of the conventional steels have been almost fully utilized. New materials and machine concepts will have to be found to do much better.

Another approach to improved efficiency is to change major components of the system. For example, the diesel locomotive is almost six times as energy efficient in operation as a steam locomotive. Theoretically, a fuel cell should be much more efficient than a heat engine. A fluorescent light is several times more efficient than a filament lamp. The transistor is much more efficient than the vacuum tube. The development of such radically new components is unpredictable, but the history of technology should lead us to expect some future successes. So far, in the past century, the raw energy input per unit of GNP has been roughly cut in half by such efforts. Another indication of this increased efficiency is illustrated in Fig. 5, which shows the effect on productivity of substituting mineral energy for human labor.

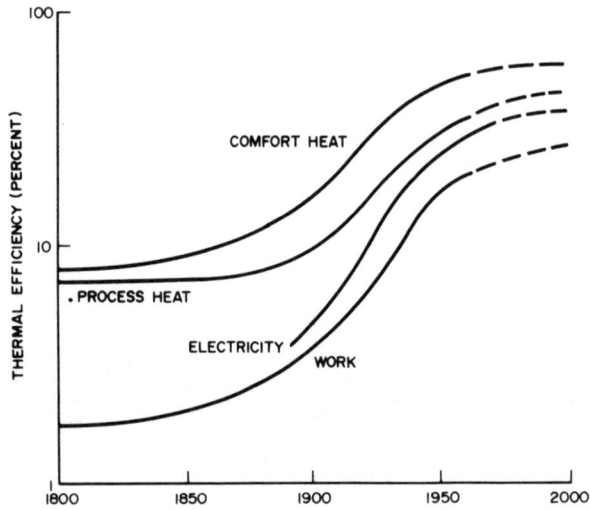

Fig. 4. Efficiency of energy converters, 1800-2000.

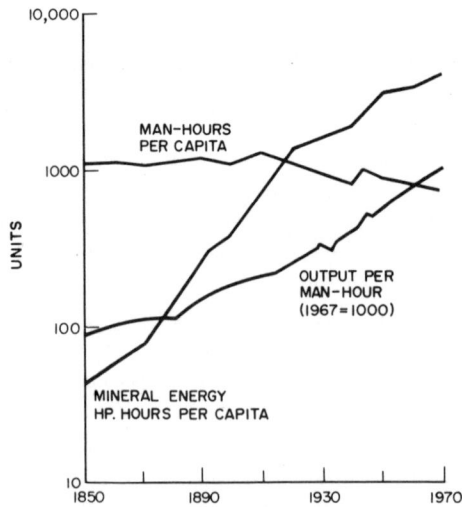

Fig. 5. US annual work energy output and man-hours, 1850-1970.

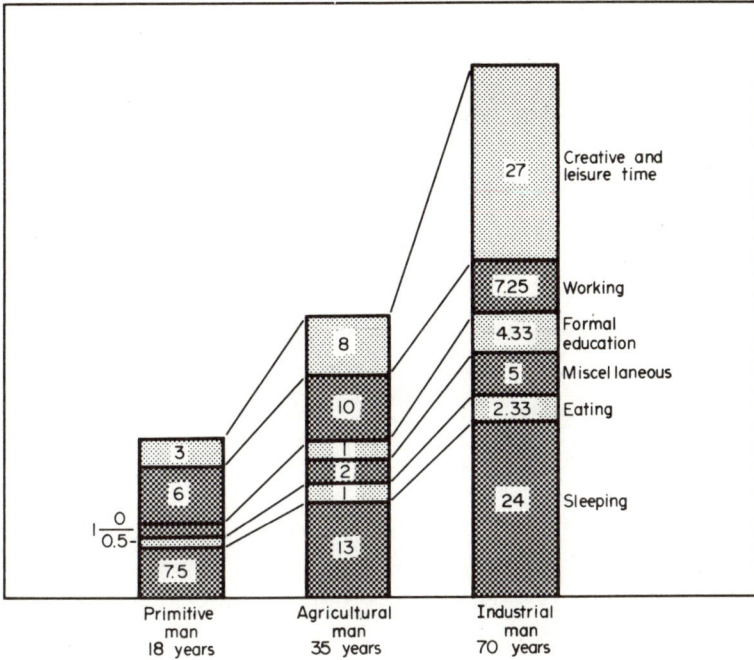

Fig. 6. Number of years spent in life activities.

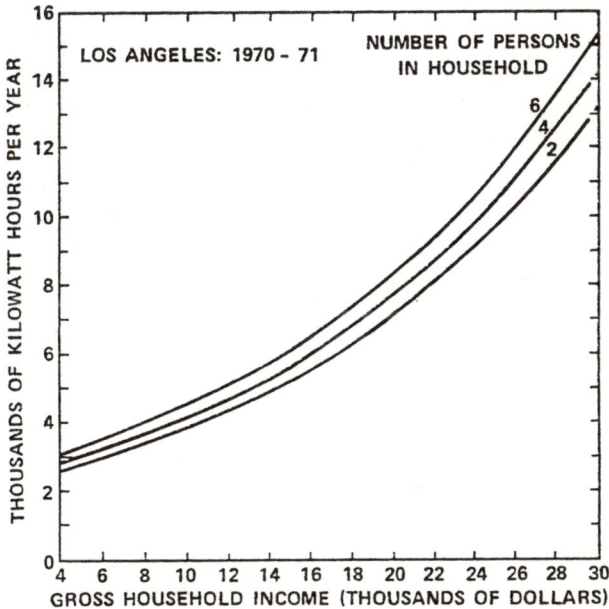

Fig. 7. Estimated average electricity consumption per household by income.

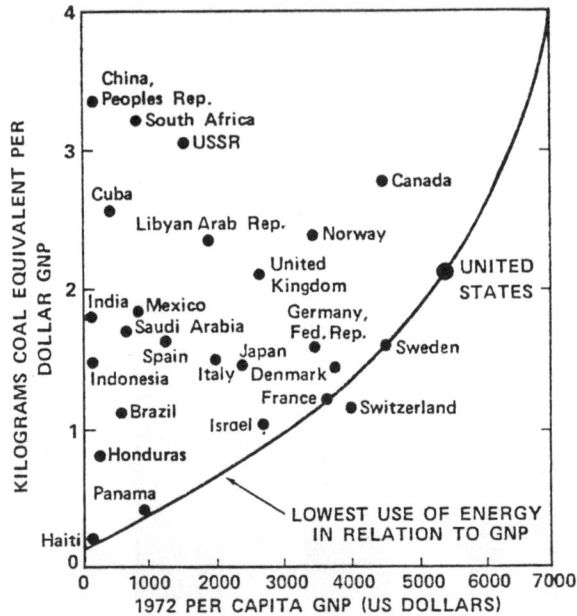

Fig. 8. World energy use.

The output per man hour has increased by over a factor of 10 since 1850 and man hours of work per capita have decreased by 30% during this same period.

Because advanced industrial countries are such large energy consumers, the productive efficiency with which they use this energy has justifiably been a matter of concern. In particular, because the US has a high per capita use of energy, it has occasionally been accused of being extravagant in this respect. The issue divides into two questions — first, is the US scale of living, as measured by the GNP per capita, really necessary for our welfare and happiness; second, is our use of energy in achieving this GNP per capita as efficient as it should be? I will not here examine the fascinating sociologic topic of how much goods and services are needed for personal welfare and happiness. It is, however, interesting to observe in Fig. 6 that the education, creative, and leisure fraction of an industrial society man's lifetime is estimated to be about 45% as compared with 25% for agricultural man, and 17% for primitive man. Further, the average life expectancy of an industrial man is twice that of agricultural man, which in turn is almost twice that of primitive man. There evidently is a relationship of these factors to the goods and services available in each society. For example, even in an industrial society, there is wide disparity in per capita energy use depending on economic income. Figure 7 illustrates a fourfold range in per capita electricity consumption for domestic use as related to income. Such a relationship may vary with climate, area, and local life style. The factors affecting these component parts are not well understood as yet.

The second question of efficiency may be examined more easily by a quantitative comparison with other countries. Such a study was made by Felix,[3] and Fig. 8

shows one of the comparisons. It would appear that overall the US is an efficient user of energy. However, because all countries differ in the constituent components of their GNP, such comparisons can be only partially informative. For example, personal services use a fraction as much energy as manufacturing for equivalent GNP. Thus, a country with a large tourism component requires a lower total energy input than one based on heavy manufacturing. This may partially explain Switzerland's position. Further, such variables as the residential space per capita, the climatic need for air-conditioning, the average distance between home and work, the availability of mass transport, all create differences in per capita energy use. Before drawing too many conclusions from this data, the fine structure of the life style and economy of each country must be better understood — as well as the relationship of each sector to energy demand.

The "lowest use" curve of Fig. 8 should not be interpreted as showing the best that can be done in regard to energy efficiency. With energy costs traditionally low, the incentive for careful husbandry has not been generally strong. Now that the cost of fuel is rising, and the social costs are being internalized in energy-system costs, the drive to more efficient use will intensify. Improved efficiency is separate but related to the parallel reduction in end-use activities which will result from these rising costs of delivered energy. It will be interesting to follow these two trends during the next several decades.

Many questions can be raised by this chart and hence only broad conclusions can be drawn. A more specific analysis is shown on Fig. 9 which shows the long-run decline in the energy use per dollar of GNP. Over the last 60 years the ratio has decreased by nearly 40%. Short-term perturbations have occurred as is evident in the 1965–1971 period. This recent rise should cause us some concern. Considering the vital link between GNP and energy, study aimed at an improved understanding of the origins of these effects would be valuable.

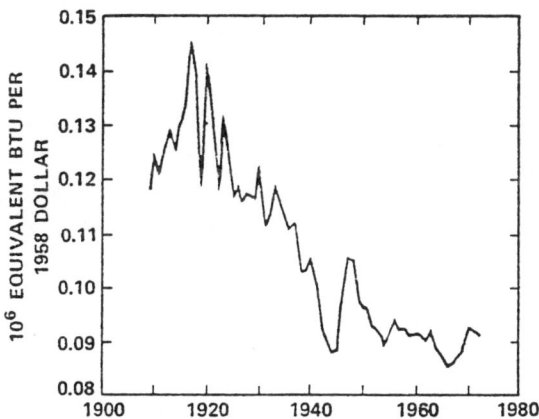

Fig. 9. Trend of energy consumption per unit of GNP in the US, 1900-1980.

II. Food, Population and Energy

In the long history of human affairs, population growth was primarily determined by the restraints of food supply. This is still true globally. Food supply in turn always has been limited by man's ability to extract it from nature. This ability has been described as passing through three major phases — as shown by Deevey[4] in Fig. 10. In the first phase, ending about 10,000 years ago, man's food supply was limited by his skill as a tool user in hunting and food gathering. There is no evidence that man used anything but his own labor in these food-collecting activities. The only energy input to the system was solar radiation absorbed by wild plants. The limitations of this process, made worse by man's social proclivity to live in small regional bands, would probably have limited the total world's population to some tens of millions.

About 10,000 years ago, Deevey suggests, man learned to domesticate animals and this relieved his food restraints. In addition to providing an increase in the ability to do work which encouraged crop cultivation, the great asset of animals was that their fuel was the natural grasses of the region, and they therefore did not necessarily have to compete with man for his own food supply. Even though the work output of domestic animals was only a few times greater than that of man, and their mechanical energy efficiency (work output compared to food input) was no better, they represented in numbers a very substantial multiplication in man's power resources. The result was the growth of an agricultural economy based on the use of domestic animals for cultivation and irrigation. Deevey suggests this might have raised the world population ceiling to many hundred million — all other factors being unchanged.

During this early agricultural period, the principal energy source was still daily solar radiation, converted through plant growth into food, and converted by man

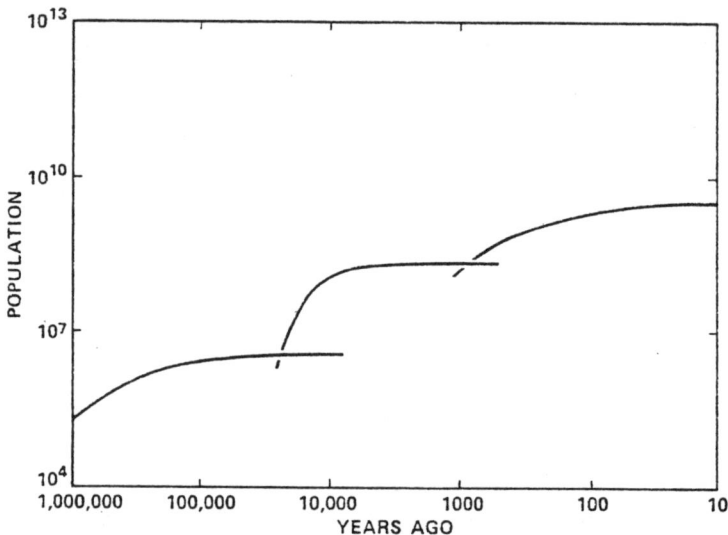

Fig. 10. World population, 1,000,000 B.C. to A.D. 1970.

and animals into work. This is a very inefficient process. The average cornfield converts about 1% of the incident solar radiation into food. Man and animals convert their food intake into work with a variable efficiency that can be as high as 25%.

So, as a source of work, grain-eating man converts up to 0.25% of the solar radiation into useful work. This is reduced by another factor of 10 if the grain is fed to cattle, and their meat used as food. Nevertheless, the use of grass-fed domestic animals for work, rather than food, represented such an increase in power resources that it expanded arable land use, created new social patterns, and stimulated population growth by raising the food ceiling. Such patterns of human existence continue to be the norm in most underdeveloped parts of the world.

During this predominantly agricultural period, man's role as a worker was subtly expanded to include the management of other work producers — domestic animals and other men (slaves notably). In addition to being a transfer machine in the chain from solar radiation to work, he now became a controller of an energy system, of which he was a part, and a goal setter as well. This was accompanied by a recognition of the opportunity provided by work devices to alter the environment, as well as to adapt to it. Although the technology of energy use for other than agriculture was in its earliest stages, the seeds for the next era were already present. Social stability and production surplus encouraged long-range investment and the civilizations of the past several millennia.

About 500 years ago, Deevey again postulates, a "scientific = industrial revolution" once more relieved the food restraint barrier, and the world population took off on a new growth curve — one we are following today, and whose upper population limit is a subject of much current speculation.[5]

The complex characteristics of the present scientific-industrial era have been analyzed from many viewpoints. With regard to energy use, however, the net effect of these developments was to make energy usefully available to man in a great abundance. The overall influence on increasing food supply has been manyfold. The work output of the individual US farmer has been multiplied by the use of farm machinery, which also released the one-quarter of the land (about 43 million acres) used to support work animals;* the amount of irrigated land has been increased by pumping; the productivity per acre has been increased by man-made fertilizer; and, very importantly, the spoilage of food between producer and consumer has been reduced by improved transportation and preservation methods. (In undeveloped countries, food spoilage may destroy as much as half that produced by farming.) In the United States, during the 40-year period from 1930 to 1970, when all these factors were fully brought to bear, the output per man-hour of farmwork increased about five times, and the productivity per acre has been roughly doubled. The US shift from a work animal to a machine agriculture is illustrated in Fig. 11. It is obvious that an energy-intensive industrial agriculture can only flourish in an industrial society, with ample energy resources. So the food problem is part of much larger issues of societal development.

It should also be recognized that such energy-intensive agricultural methods depend on the addition of stored solar energy (fossil fuels chiefly) to the direct

*This is about equal to the arable land in all France.

CURRENT ISSUES IN ENERGY

Horses & Mules (millions)   Tractors (millions)

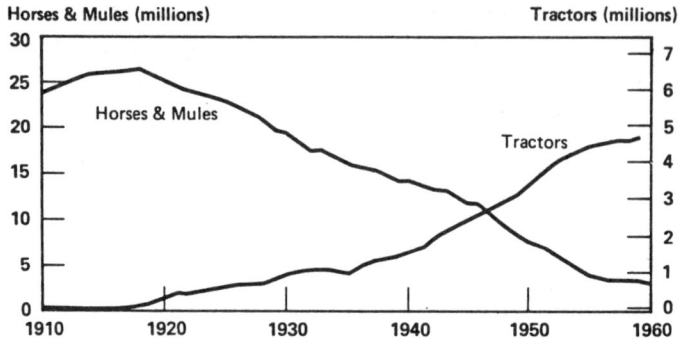

Fig. 11. Machines versus animal use in US agriculture, 1910-1960.

incident solar radiation in increasing the yield per acre or yield per man-hour of work. In the case of corn, it has been calculated[6] that the corn yield per acre is increased roughly threefold by a fossil-fuel energy supplement equivalent to about 11% of the solar energy absorbed directly by the corn plant. The efficiency of food production using supplemental energy input decreases as the yields go up.[7] In primitive agriculture, the ratio of food energy available to non-solar energy input (man-work) was about 16. In the work animal agriculture this ratio dipped to 3–6 times the total work input; and in energy-intensive agriculture, the ratio may range from 0.3 to 2 depending on the food product.[8] Intensive agriculture is not energetically efficient, although it is effective for increasing food production per unit of arable land. However, there is probably an upper limit close to the best output of present methods.

In the US, about 12% of the total national energy consumption is required for its food cycle. This 12% consists roughly of 3% entering directly into agriculture, 4% into food processing, 2% into transportation and merchandising, and 3% into household preparation. So 3% of our total energy consumption permits US agriculture not only to feed the nation, but also to grow a substantial surplus for others. On the timely issue of petroleum, agriculture accounts for 3% of this fuel use. The food cycle also accounts for about one-fifth of US electricity consumption. It is evident that the three principal parameters of manpower, arable land, and fuel may be varied to provide the trade-offs appropriate for each society. Intensive energy use can certainly raise food production. However, this also has limitations. If all the proven oil reserves were used to fuel intensive agriculture, then a 4 billion world population could be given an average US diet for 29 years; or about 50 years on a corn grain diet.[6] These numbers, even if very rough, illustrate the magnitude of the world food problem.

The historical exponential growth in world population, and the obvious need for population control, has not visibly resulted in a substantial decrease in birth rate except in developed countries. By the year 2000, the industrial countries are expected to increase their population by one-third, whereas the non-industrial countries are expected to double their population. One can hypothesize that the reduced need for manpower in a society where less costly machine labor is available to take its place produces major social forces to reduce the desired family size. In contrast, the capacity of a non-industrial agriculture to absorb human labor is very great. Presumably in an industrial society, the perceived potential benefits of children to the average family unit are intuitively analyzed and compared with their potential burden on the family. First, a reduction in childhood mortality, arising from food and material availability and medical aids, decreases the number of births required to achieve a given number of adults. In many underdeveloped areas, less than half of all children born live to become adults. Second, the contribution of children to the family is less in an industrial society. To quote Coale (reference 5, p. 51):

"In an urban, industrial society the family is no longer the main locus of economic activity, nor are children the expected means of support in old age. In an agrarian, preindustrial society, on the other hand, the family is a basic economic unit and sons are a form of social security. Moreover, in the less developed countries the costs of raising and educating a child are minimal; indeed, a child may contribute to the welfare of the family from an early age. In the industrial society child labor is prohibited, education is compulsory and it often extends through adolescence. These conditions conspire to discourage couples in the more developed countries from having large families, whereas in long-established agrarian societies social norms supporting childbearing tend to be perpetuated."

Third, as the average productive life span of an adult increases, there is a decrease in the birth rate needed to provide the same man-hours of adult work.

The above factors all enter into the intuitive perception of the cost-benefit aspects of family size which establish societal patterns and traditions. Changing such patterns is a slow process because it takes a long time for changing trends in the above factors to become visible. Thus, sudden technical interventions that substantially reduce death rates can cause transient population surges until their effects are perceived (see Coale, ref. 5, p. 49).

The key point to be made here is that the substitution of work machines for manpower, and the accompanying expansion of material welfare, can provide a fundamental motivation for reduced family size. This could be especially effective because the process will inevitably be slow enough to permit its impact to be perceived. For example, the birth rate in the US has been generally dropping for the past 150 years (as shown in Fig. 12), roughly the same amount of time it took to make the transition from an agricultural to an industrial society. The death rate has also dropped, but the net difference is trending to a population equilibrium. It is important to observe that this trend has long been underway, without an overt public campaign for population control or the planned promotion of birth-control techniques. Thus, industrialization may be one of the means which effectively reduces the birth rate and, like the transistor radio, could transcend all cultures and customs.

The potential task of providing sufficient energy to meet such a worldwide goal is itself formidable. The rough numbers are descriptive of the problem. Assuming that half the US average energy consumption could provide a life style similar to that in the US now (by means of future increased efficiency and conservation), and assuming the world population only doubles before equilibrium is reached,

Note: Except for 1860–70, the data for years prior to 1910 are charted at intervals of a decade; for years after 1910 the chart reflects yearly data.

Fig. 12. US birthrate, 1800-1974.

then the annual world energy need will be almost ten times its present consumption.

The limitations on the quantity of stored fossil fuels and their recoverability have been discussed elsewhere. Regardless of the range of such projections, it is quickly apparent that fossil fuels are unlikely to be able to supply an industrialized world very long. Of the methods available now, this can only be done by using nuclear fission power, in particular by the fast breeder. If future technology permits, nuclear fusion will do it. The most challenging resource, of course, is the use of continuous power from solar radiation — a technology which is available now for low-temperature uses, but which is yet much too capital demanding for the high-temperature purposes necessary to energize work machines.

In analyzing the alternative approaches to massive increases in world energy production, the environmental consequences must be carefully projected. The pragmatic limitations of the biosphere with regard to the by-products of energy production are very poorly understood as yet. Much more comprehensive research is needed in this field. In theory, all effluents from energy systems, except the ultimate low-temperature heat output, can be controlled and contained. In practice, the constraints of nature, technology, and economics make the theoretical ideal difficult to approach. We will need to know much more about the trade-offs that will eventually have to be made. The ultimate waste-heat issue probably can be handled acceptably if the world energy increase is only a factor of 10. The heat-disposal problem at this level is likely to be more one of local energy density than one of global significance. In fact, this low-grade heat may eventually be a constructively useful resource in many parts of the world — for district heating, agriculture, and mariculture.

Clearly, all energy production methods must be explored to provide acceptable options for meeting such a future worldwide objective. Unfortunately, even if energy were to become easily available to all, industrializing the world will be a slow process — with the energy availability only one of many factors. Nevertheless, energy availability can be accelerated by technology. It is one of the few technological steps the developed countries can undertake now to help the world's future welfare.

Conclusion

It is evident from this limited survey that human society has a complex energy ecology woven into its fabric. As with the ecology of biologic systems, the ecology of energy systems has both direct and subtle relationships between its internal components, the natural environment, and the societal environment. Any one change in this total system of energy relationships affects all its component parts.

Our ignorance of these matters has led to many simplistic notions concerning the role of energy on the quality of life, ranging from the poetically idyllic notion that the less energy used the better the life, to the opposite extreme of the need for unlimited abundance. Both extremes have little evidence to support their validity. A grain of truth does not make a whole doctrine.

Observable social groups with very low energy supplements are frequently on starvation's edge, ridden with malnutrition, endemic disease, and physical misery. There is no mass migration into such deprived societies. It has been estimated that one-fourth of the world now lives on this edge of survival. Given a free choice, such groups avidly seek more energy inputs. It is clear from man's history, and from the visible living patterns in the range of societies worldwide, that a significant level of inanimate energy to supplement human labor does, by almost any humanistic measure, improve the quality of life. At the other extreme, the affluent members of our most industrialized societies are apparently satisfied with a per capita energy use only a few times greater than the average in those industrialized societies.

The important questions for the future relate to the magnitude and form of energy supply, and the value systems used to make related trade-off judgements. In more pragmatic terms, the field of energy ecology requires much more insightful study and analysis of the very many factors involved. To list but a few — in the technologic domain the issues of performance efficiency and energy system effluents and impacts; in the economic domain, the structural changes determined by energy system costs and options; in the sociologic domain, the effects of life style, population distribution, and public health and welfare; in the political domain, the management institutions for energy systems, and their relation to national welfare, foreign policy, and the survival and welfare of future generations.

These are obviously complex subjects, only superficially understood as yet. Under these circumstances, sweeping approaches to the current problems we face nationally in the energy field have dubious validity. Simplistic general objectives, such as "reducing energy consumption" have as little constructive meaning as advising the nation to "reduce food consumption", and may have equally dangerous consequences. Selective objectives based on system insights and trade-offs must form the basis of our national decisions in the field of energy ecology. Eventually, the most developed societies will be those that achieve a high average level of individual welfare with a minimum use of all resources — of which energy is but one, although a very important one. The challenge we face is to approach this goal with a minimum number of errors and false starts. Even assuming we have a national consensus on our societal objectives, in developing national programs we need causal insights rather than shallow cliches. The anatomy of our society deserves the scalpel rather than the axe, and this is especially true as we undertake to administratively modify the key energy systems that maintain it.

The following quotation from the December 1922 *Scientific American* illustrates the hazard of assuming an unchanging technological base in long-range projections:

> "According to Eugene Davenport, Dean of the Illinois College of Agriculture, the greatest future need of American agriculture is a fundamental national policy. One of our leading statisticians estimates that a century from now our population will amount to more than 225,000,000 people. The prophecy is startling because it suggests possible hunger and even famine as our future. At present, with less than half these numbers, our food production is only about equal to our domestic consumption. Unless we institute very revolutionary practices to enhance production, we may look for it to fall behind. Present indications are that we are rapidly slipping into the class of food-importing nations. This means that unless we are able to reverse the tide, we must readjust our social, economic and industrial organizations to accord with this new condition."

References

1. S. Schurr, B. Netschert *et al.*, *Energy in the American Economy, 1850–1975*, The Johns Hopkins Univ. Press, 1960.
2. R. G. Lillard, *The Great Forest*, Knopf, 1948.
3. Fremont Felix, *Electrical World*, 1 Dec. 1974.
4. E. S. Deevey, The human population, *Scientific American*, p. 198, Sept. 1960.
5. A. Coale, The history of the human population, *Scientific American*, Sept. 1974.
6. D. Pimentel, L. Hurd *et al.*, Food production and the energy crisis, *Science*, 2 Nov. 1973.
7. J. Steinhart and C. Steinhart, Energy use in the US food system, *Science*, 19 April 1974.
8. G. Borgstrom, Food, feed and energy, *Swedish Academy of Science*, Dec. 1973.

8

Energy Use: An Interregional Analysis with Implications for International Comparisons*

The ratio of economic activity to total energy consumed in Btu's — the Gross Domestic Product/energy ratio (GDP/E) — has been used extensively to symbolize a relationship between an economic surrogate for the so-called "well-being" of societies and the consumption of energy inputs. Studies have been conducted by Schurr et al.,[18] Darmstadter et al.,[7,8] Schipper and Lichtenberg,[17] Goen and White,[9] Netschert,[13] Boretsky et al.,[5] and Berndt and Wood,[1,2] to name a few authors. This GDP/E ratio has been hailed by the popular press as a measure of the efficiency with which nations use energy fuels. More, the comparisons with foreign countries have been used to indicate the potential for energy conservation in the US.

For example, in a recent *Business Week* article[22] it was pointed out that West Germany's industrial sector uses 38% less energy per unit of output than American industry; Sweden uses 40% less energy to produce each dollar of GDP. On the consumption side, this article stated that Swedish homes burn less fuel than American homes, that Swedish cars get 24 miles per gallon, and that the Swedish drive less because they have an excellent transportation system. Aside from the uncertain validity of these numbers, the implication is that based on the example of these countries, achieving conservation savings in the US of 30–40% should be readily possible. Such international lists of "they use energy better than us" suggest to readers that the US is sloppy in its use of energy.

The merit of these contentions depends on the interpretation of the meaning and significance of GDP/E ratios. That is, what information does the GDP/E ratio provide? Can GDP/E comparisons measure the relative "efficiency" of energy use? And based on these relative measures, is there some information that would tell us how, or in what way, we should redirect our resources to better uses? Allied to all of this, can such ratios aid us in making projections for future energy planning? It is to these questions that this paper is addressed.

---

*Presented at the Resources for the Future, International Energy Consumption Comparisons Workshop, Washington, DC, 15–16 September 1977.

Problems in Measuring Economic Efficiency

In mechanics, work output divided by energy input is defined as a measure of efficiency, and the higher the efficiency of a system, the closer efficiency is to unity. We all know this. But the world of economics is not the world of theoretical or applied mechanics. As is demonstrated below, GDP divided by energy does not measure how efficient the US or some other society is in its use of energy. In the generation of some GDP, energy is only one factor of production. Capital, labor, and materials are also factor inputs. And societies (industries and firms) combine these factors in different ways to produce outputs. Just what combinations evolve depends on several factors, one of which is prices. If one society shows a high GDP/E ratio relative to another, this does not mean that some given society is more efficient in its energy use. Certainly, there are alternative interpretations, e.g. that the allocation of factor inputs is different for each country. Their resource base, technology, industrial mix, and so on are different. Or should the price of energy be relatively high, one country may substitute labor for energy; another, capital for energy.* This is not to say that GDP/E is not a useful parameter in energy analysis. In any given region, with only slowly changing factors of production, this ratio does permit useful trend analysis and can provide insight to the regional energy-use patterns.

To illustrate the inadequacies of using GDP/E ratios to measure the relative "efficiencies" of countries, Table 1 provides a combination list of alternative economic measures. For example, economic value could be combined in ratio form with capital, labor, energy, materials, or technology; alternatively, physical output could be combined with any one of the resource variables, and so on. The ratios generated would likely be different for each country, but no conclusions could be drawn as to the relative "efficiencies" of capital, labor, energy, materials, or technology employment from such ratios. In fact, such ratios omit the key elements of the existing industrial infrastructure and the organization and management of productive units.

TABLE 1
Alternative Economic Measures

Goods and services		Resources
Physical output		Capital
Economic value		Labor
Physical consumption	Vs.	Energy
Utility		Materials
		Technology

---

*Griffin and Gregory,[10] employing international cross-sectional data, suggest that, in the long-run, capital and labor are substitutable for energy.

If the goal is to measure "economic efficiency", then one approach is to define efficiency as the ratio

$$\frac{\text{Actual economic output}}{\text{Maximum economic output}}$$

for a given mix of resources. Here a ratio of unity defines optimum efficiency and is achieved when the value of each (marginal) productive service equals its alternative cost.* Obviously, if the alternative cost is less than the value of the marginal product, a unit of the productive service will produce more in this use than elsewhere, and output is not at a maximum. If the alternative cost exceeds the value of the marginal product in any use, a, unit of the productive service will produce more elsewhere, and output is not at a maximum. Optimum efficiency, a ratio of unity, is achieved when the value of the marginal product of each productive service equals its alternative cost. A difference between alternative cost and the value of the marginal product (say, for energy) in any firm or industry is proof of inefficiency, and the magnitude of the difference is a clue to the extent of the inefficiency.

We suspect that the condition for maximum output (measured in terms of the prices consumers are willing to pay) is being fulfilled in free market Western societies today. That is, energy is being employed up to the point where its marginal product just equals its alternative cost. If this is the case, then Sweden, West Germany, and the US have equal economic efficiency in their use of energy.

Interregional Comparisons

Another tack to providing insight into the problems of using GDP/E ratios for comparative purposes is to make an interregional analysis. This we have done in our "Proficiency" paper[19] where we have given a look at the differences in economic output and energy use in each State of the US. What did we learn from this exercise?

As expected, we found a wide range of energy use and aggregate economic levels between the States of the US, explained in part, by differences in the mix of economic activities and from the differences in energy consumption per unit output of these activities. Each State has its own economic characteristics which have evolved from its location, natural resource base, population (and concomitant labor force), and other parameters. But more, each State has developed and grown as part of the whole US economic structure a structure based on the relatively free flow of resources (including energy), labor, materials, semi-finished and finished goods between States. That is, through the division of economic activities (or specialization), the benefits from inter-state trade have become large for society as a whole.

Figure 1 discloses the wide range of energy use and economic activity that existed between the States in 1971. However, to argue and compare the efficiency of States by their respective Gross State Product (GSP)/E ratios does not mean

---

*Alternative (or opportunity) cost is the maximum amount that any productive service would produce in any commodity.[20]

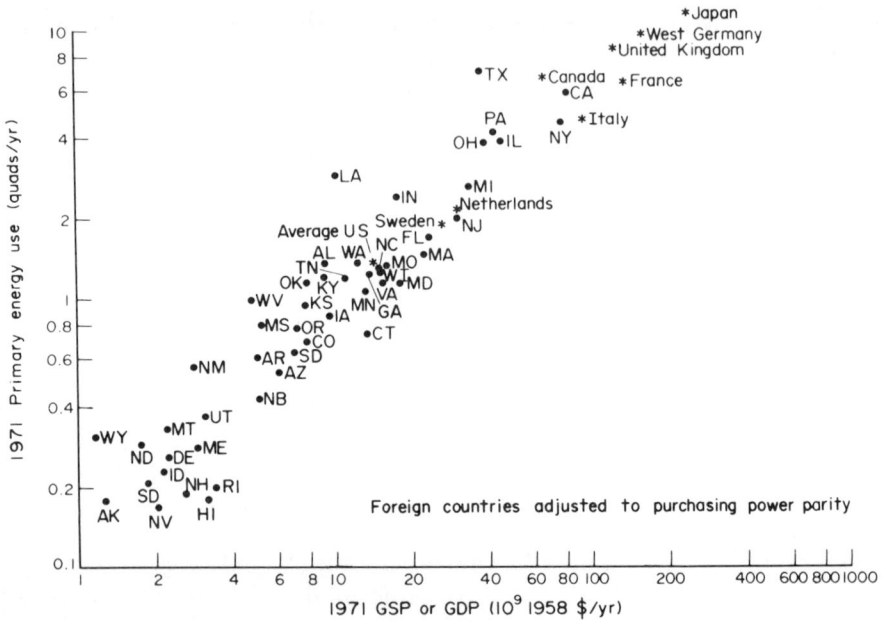

Fig. 1. Gross product vs. energy use, 1971.

one State is more energy efficient than another. On the contrary, each State is engaged in economic activities associated with different energy requirements. Compare, for example, Wyoming and Vermont, Louisiana and Connecticut, or Texas and New Jersey. A cross-sectional sample of State data points for 1971 is provided in Table 2. Both Wyoming and Vermont have similar Gross State Products, but Vermont uses far less energy than Wyoming. The GSP/E ratios tell us so. Is Vermont more energy efficient than Wyoming, than Louisiana or Texas?

TABLE 2
Sample Data: State Pairs, 1971

	Wyoming	Vermont	Louisiana	Connecticut	Texas	New Jersey
GSP* ($ billions/yr)	1.3	1.5	10.9	14.3	40.4	31.2
Energy use (quads**/yr)	0.33	0.084	3.07	0.78	7.55	1.98
GSP/E ($/million Btu)	3.9	17.5	3.6	18.4	5.4	15.8

$$*GSP = \frac{\text{State personal income}}{\text{National personal income}} \times GDP. \text{ Methodology from reference 11.}$$

**quad = 10^{15} Btu.

The corollary between States and inter-country comparisons is an interesting one in that GDP/E comparisons are just as involved as they are for States. Can we compare Sweden, West Germany, or Japan with the US using GDP/E ratios?

Perhaps a more likely comparison would be between Florida and Sweden; the Netherlands with New Jersey; Italy with New York; Canada with California. None of the comparisons, of course, are really justified using economic/energy ratios. Each country is as different as each State, and there is a certain absurdity in trying to compare energy use in one to another or in concluding that one is more energy efficient than another.

Issues in International Comparisons

We have all seen, in the publications cited, comparative international GDP/E tables or plots, where GDP in foreign currencies is converted to US dollars using exchange rates, or the more meaningful purchasing-power parity indexes,[12] and deflated to constant dollars. Additionally, energy and national output may be on a per capita basis. Sweden, the Netherlands, West Germany, the United Kingdom, France, Italy, and Japan, to name a few countries, all have higher ratios than the US. And it is assumed, or explicitly stated, that these countries use energy more "efficiently" than the US. As has been pointed out, this, of course, is misleading. But more, there are a number of issues that need to be raised from this type of analysis — questions of whether or not such analysis holds any constructive meaning for the US.

Production vs. Consumption

There appears to be some confusion in the literature as to what is being measured: the *production* of goods and services and the relative "efficiency" of energy input to the creation of any given GDP bundle, or the relative *consumption* of goods and services and the energy embedded therein. That is, energy use in any given economy occurs mainly in those intermediate activities that produce goods and services that are measured by GDP. The remainder of the energy (whose price is included in GDP) is used to satisfy final demand, i.e. purchases by final consumers for such things as home-heating fuels, residential electricity, gasoline, and so on. Obviously, the split between energy for final demand and for economic output is implicit in the GDP/E ratios. For example, from Table 3, in 1972, total primary energy used in the US was 72.9 quads. Of this, 45.2 quads or 62% of the total were employed in the intermediate production process; 27.7 quads were consumed as final demand. This split, of course, is important, and the breakdown of how energy is being consumed in the final-demand category and the various allocations being made in the industrial production sectors need to be identified. And they will vary from society to society. But simple GDP/E ratios do not tell us this, and if the intent is to identify the relationship between energy use and output (the "efficiencies" of energy to the creation of some level of GDP), then we are taking about *production*. Here the mix of industry, technology, price, resources, and so on need to be considered. Alternatively, if the final *consumption* side of energy is to be considered, besides price and income, the mix of goods and services available and their energy intensities, demographic characteristics, lifestyles, climate, and so on will need to be reviewed. Confusing the two can result in incorrect interpretations.

TABLE 3
Primary Energy Use in the US, 1972, 10^{15} Btu/Yr

	Economic ouput activities	Final demand	Total primary energy
Coal	12.2	2.2 (including 1.8 exported)	14.4
Oil	15.4	18.2	33.6
Gas	15.3	6.1	21.4
Hydroelectric	1.9	1.0	2.9
Nuclear	0.4	0.2	0.6
Total	45.2	27.7	72.9

Allied to the issue of production vs. consumption is the question of embodied energy, i.e. whether or not the final consumers of a product should be charged for its embodied energy (i.e. all the energy inputs required for its production) or whether that society (or state) which receives the value added in the production process should be charged directly for the energy it employs. It would appear that if one is focusing on production, or the relationship in each country between the economic value generated (GDP) and the energy required to create this economic value, it makes little economic sense for the energy embodied in the final product to be charged to the country that consumes the final product. That is, energy, in some combination with capital, labor, and materials, was employed to help generate some level of GDP, effective demand, and employment. Thus, the employment of Btu's contributes to the so-called "well-being" of any society. And in the I/O relationship of GDP/E, subtracting out the energy employed for products exported obviously incorrectly biases the ratio upwards. Similarly, adding embodied energy in imports incorrectly biases the ratio downward. In comparative analysis, such interpretation would be misleading if the objective is to compare "efficiencies" or "effectiveness" of energy usage to the economic values generated.

But, if one is asking the question of how much energy a nation is consuming, not in terms of production, but in terms of worldwide contributions of energy to consumption, then embodied energy may be considered pertinent. But the intentions of such analysis must be clearly stated, namely, to show the additional amounts of energy consumed from traded goods. But then, comparative analysis, and any presumption from this analysis, that concludes that foreign practices and consumption would be profitable to explore would be misleading if "efficiencies" or "effectiveness" is the intent. That is, the analysis would be in terms of consumption patterns, and not the productivity of energy use.

An extremely simple example illustrating the consumption/production viewpoint is provided in Fig. 2. Here countries A and B each consume one dollar's worth of steel and one dollar's worth of watches. The energy intensity of steel is 5 units of energy per dollar; the intensity of watches, 0.5 unit of energy per dollar. It is assumed that the import by country B of one dollar's worth of steel is equivalent to the import of 5 units of energy. Similarly, import of one dollar's worth of watches by country A is equivalent to the import of 0.5 unit of energy.

Using the production viewpoint with this example, the GDP/energy ratios when not adjusted for imports and exports are 0.2 for country A, and 2 for country B. The philosophy behind this is contained in the question: How much output (GDP) does a country produce with a given amount of energy? When adjustments are made for imported embodied energy, the GDP/energy ratio of 0.36 is the same for both countries A and B. If the consumption viewpoint is employed, both countries look equally energy consuming; alternatively, given the production viewpoint, country B would appear more energy "efficient" than country A. Neither approach serves to tell us whether one country is more or less "economically efficient" than the other.

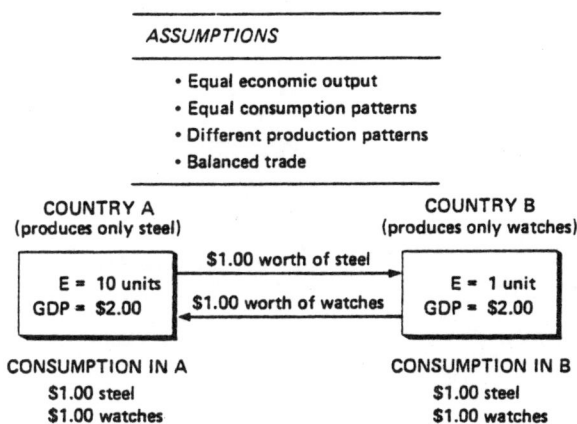

```
┌─────────────────────────────────┐
│  ASSUMPTIONS                     │
├─────────────────────────────────┤
│  • Equal economic output         │
│  • Equal consumption patterns    │
│  • Different production patterns  │
│  • Balanced trade                │
└─────────────────────────────────┘
```

COUNTRY A   COUNTRY B
(produces only steel)   (produces only watches)

```
┌──────────────┐   $1.00 worth of steel   ┌──────────────┐
│  E = 10 units │ ──────────────────────→  │  E = 1 unit   │
│  GDP = $2.00  │   $1.00 worth of watches │  GDP = $2.00  │
│              │ ←──────────────────────   │              │
└──────────────┘                          └──────────────┘
```

CONSUMPTION IN A   CONSUMPTION IN B
$1.00 steel $1.00 steel
$1.00 watches $1.00 watches

Fig. 2. Embodied energy. Consumption/Production viewpoint.

Industry Mix

Another issue is that of industry mix. In our "Proficiency" paper,[19] we noted the tremendous variations in value added and energy use for some eighty-seven sectors in the US economy.[21] And it was shown that heavy industries such as iron, steel manufacturing, chemicals, oil and gas extraction, petroleum refining and paper require substantially more energy for a specific level of value added than other economic activities such as printing and publishing, finance, and insurance. Figure 3 provides a broad brush picture of the economic activities in the US for the year 1972. And those States such as Texas and Louisiana that are noted for a large proportion of energy-intensive industries, e.g. petroleum refining and petrochemicals, and Indiana, Ohio, and Pennsylvania where steel manufacturing is prominent all have much lower GSP/E ratios than those States that provide services such as New York which specializes in banking, insurance, and real estate.

Another interesting feature is shown in Fig. 4 in that those States that have a relatively low GSP/E ratio are predominantly energy exporters. Most of these

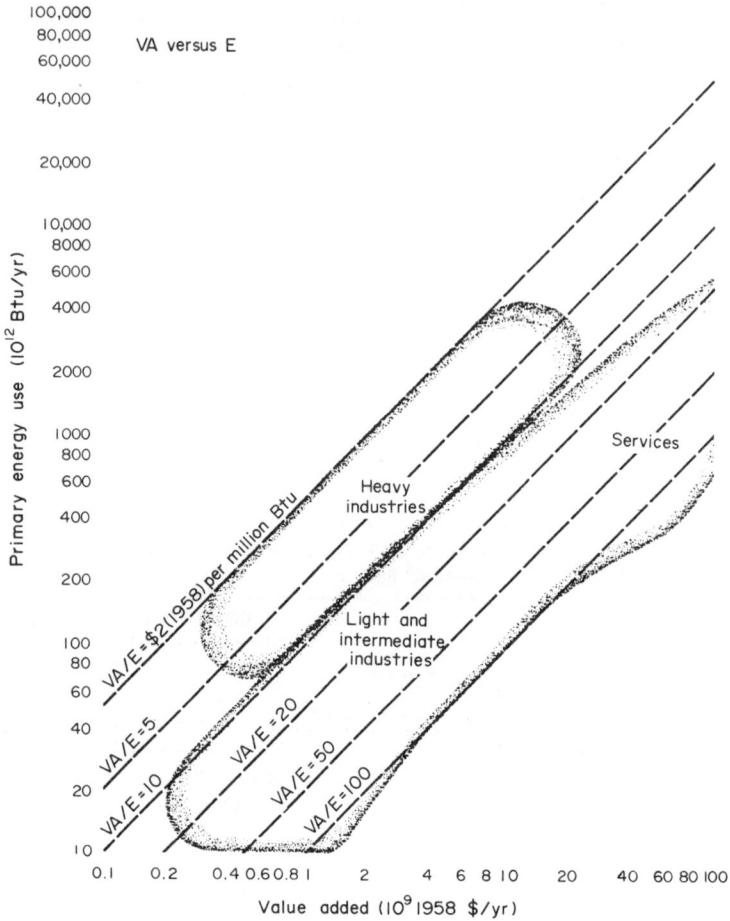

Fig. 3. Economic activities in the US 1972.

energy-source States not only have installed energy-intensive conversion facilities such as oil and gas-extraction plants, but energy-intensive satellite industries have located and developed in nearby areas. Similarly, in looking at the GDP/E ratio internationally versus the dependence on energy imports, Fig. 5 for selected countries[8] shows a similar relationship.

This relationship between the GDP/E ratio and dependence on energy imports is an interesting one in that higher ratios appear to be associated with greater energy dependence. Part of the explanation lies in the fact that those societies that have become increasingly dependent on energy imports have orchestrated policy directed to minimize this dependence, e.g. tax policies directed to raise energy price levels, particularly on petroleum products. These higher prices not only restrict the quantities being consumed, but act as a mechanism to allocate energy to more efficient uses. And, in part, these higher prices will serve to shift the industrial base away from energy-intensive-type industries to less energy-

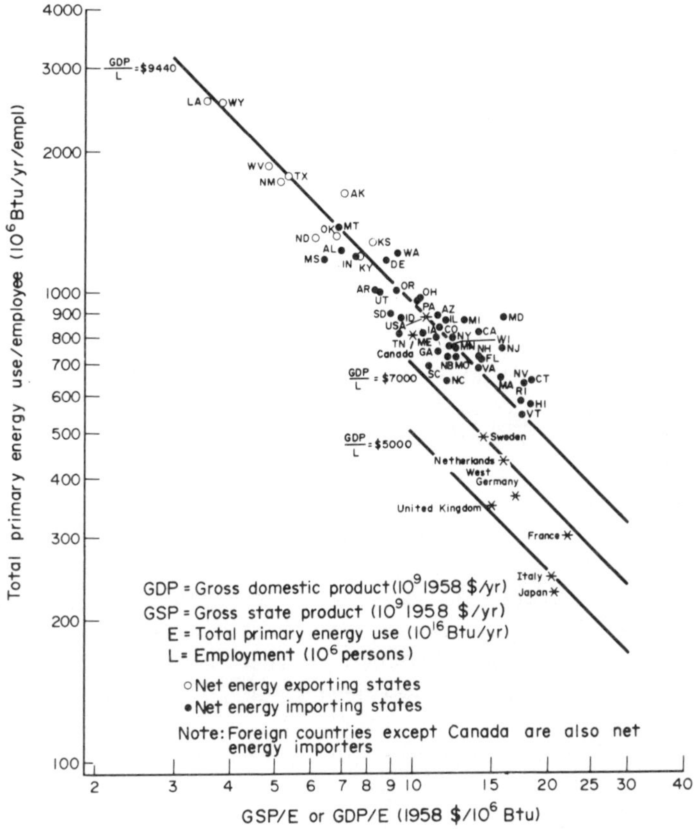

Fig. 4. GSP/E or GDP/E vs. E/L for US and foreign countries, 1971.

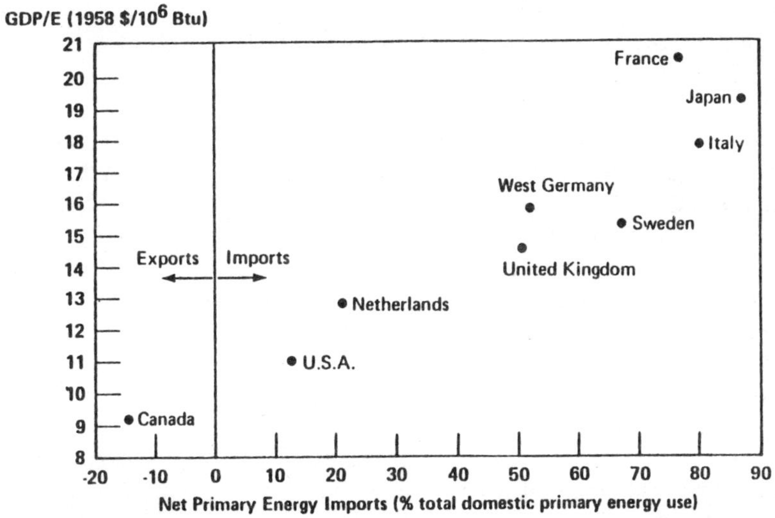

Fig. 5. GDP/E vs. dependence on energy imports for 1972.

intensive types, or from those industries with low value-added contributions to high value-added sectors.

The point is that the industrial mix for each society is likely to be different, and because of these differences, the quantities of energy employed for production will be different. And if comparable data could be found for the major Western societies, the differences in industrial mix would become clearer, and help to explain variations in energy use.

Additionally, comparisons of aggregate data for specific industries can be misleading. As examples, we can point to two major industries: petroleum refining and steelmaking. The summary of the SRI report[9] indicates that West German energy use is 67% of that of the US per ton of petroleum products manufactured. That is to say that the average US refinery uses 11.2% of the crude oil as refinery fuel, whereas West Germany refineries use only 7.5% of the crude oil. This is explained to a large degree by the fact that the average US refinery produces substantially more gasoline, kerosene and jet fuel from a barrel of crude oil. To achieve this greater production of light products, it is necessary to use fuel to provide the process heat necessary to crack the heavy hydrocarbons in the fuel oil portion of the crude oil. For example, catalytic cracking and hydro cracking capacity in the US is 36% of crude-oil capacity. In West Germany, the comparable figure is only 4%. On a first-cut basis, it would seem inappropriate to compare refinery fuel utilization per unit of refined product because of the substantially different product mix. And likely, refineries use fuel as efficiently in the US as in West Germany.

Concerning steel manufacturing, after mining, milling, and concentration, iron ore is sent to a blast furnace which derives most of its heat from the energy in the coke that is added. The energy requirement in the US in 1972 was about 42.6 million Btu/short ton of pig iron. Ore concentrations in the US are about 50–55% iron, whereas in Sweden, ores are richer in iron (60–65% Fe). The richer ore probably has less requirement for energy. Thereafter, pig iron is mixed with scrap steel (derived from both outside purchase and internal mill recycling) and sent to three types of secondary processing: basic oxygen, open hearth, and electric furnace, the product of which is raw steel. The choice of secondary processing is dependent not only on economics but also on the type of steel to be made. Sweden has the reputation of being the "jeweller" of the steel industry. That is, they produce, on the average, a higher value added product. This is reflected in figures, calculated from Schipper.[17]

Value Added/Primary Energy

(1971 $/million Btu)

	US	Sweden
Basic steel	3.4	6.8
Primary metals (incl. basic steel)	4.5	5.3

But these figures do not reflect feedstock variations, processing variations, and differing product mixes and certainly they do not allow one to estimate conservation potential.

The point of all this is that as one subdivides the aggregate figures for each industry, it becomes clear that each industrial sector is quite complex and quick conclusions are not attainable. It is necessary to examine in detail feedstock quality, product mix, specifications and the use of recycle materials all in relationship to current and anticipated costs of capital, labor, energy, and materials in order to understand international differences.

Lifestyles

Another important issue when comparing GDP/E ratios is that of lifestyles. Here, of course, we are referring to the consumption side of GDP, and not the generation of product. Certainly, the way we live, our "habits" of consumption, can be of importance when the conservation of energy becomes an issue.

In the US itself, the South has much lower incomes than the Northeast and the West Coast. States such as Mississippi have scarcely attained per capita incomes equal to those reached in Pennsylvania and New York in 1925 or those found in France today. In other words, per capita incomes vary from State to State. But per capita incomes (and even the prices of energy) may not sufficiently serve for us to understand why there are differing consumption patterns of energy. In addition, social, political, other economic, and environmental conditions also vary from State to State, or from region to region, i.e. lifestyles, or indexes of "the quality of life", may help in interpreting and explaining differences in energy consumption. And these lifestyles, the social environment in which people live, should be subject to analysis in international energy comparisons studies.

Two recent studies by the Midwest Research Institute[3,4] show that the quality of life varies from State to State and from Standard Metropolitan Statistical Area to Standard Metropolitan Statistical Area (SMSA). Based on a composite index of social, economic, political, and environmental variables, California ranked number 1, Colorado number 2, Alabama number 50. On an income per capita basis, California ranked number 5, Colorado number 21, and Alabama number 44. In other words, income per capita may not serve as a sufficient measure to explain the "well-being" of people and the overall environment in which they live. For an overall index for large SMSAs, Portland, Oregon, ranked number 1; Jersey City, New Jersey, number 65.

Based on these differences in lifestyles or the quality of life in the US, we would expect to find differences on an international basis. One international study[16] supports this contention. In this study a composite index was developed that included, for example, the number of telephones, cars, radios, and television sets in use and the share of expenditures for food, drink, clothing, rent, recreation and energy in personal consumption, and so on for Holland, Luxemburg, Belgium, West Germany, Austria, and Italy. The index quantified lifestyles in these EEC countries. Not surprising, wide differences in lifestyles were supported by this historical evidence, i.e. the people in Holland have different lifestyles from those in Italy or Austria.

A more recent publication[6] shows the relationship between man's use of energy and the social fabric or lifestyle of that society for different countries of the world; it too points out that people do live differently and employ energy in uses based on that lifestyle. Additionally, Cook points out that changes in lifestyles are partially brought about through increased uses of energy, i.e. there is a positive relation between energy and the way people live.

The point is that in looking at and comparing energy use in societies, lifestyles are an important parameter, too often neglected in analytical efforts. And what is required in international energy-comparative studies is to give a hard look at how people live, because energy consumption is going to be reflected in any given society's lifestyle. The task is not an easy one, as the quality of life or lifestyles mean different things to different people and, at present, no consensus exists as to what it is or what it means. Yet a consensus does exist regarding its importance as related to energy, and for this reason it needs to be explored further. And in its exploration, international differences in energy consumption may become clearer.

Maximizing End Function and Minimizing Resource Utilization

Interregional comparisons have two useful objectives. The most important is their continuing contribution to understanding energy flow and human welfare. A second objective is based on the hope that such studies will disclose regional conservation potential from available aggregate measures. For reasons stated above, this is not possible.

As Fig. 6 shows, the efficiency of energy use (defined by the relative primary fuel requirement to produce a given end function) is the product of a chain of individual process efficiencies. In each process of the resource-to-end function sequence, external factors do play a role. When these external factors are significantly different among countries, the use of international data on a sector basis is misleading.

Process	Factors
Resource extraction and transportation	Quality
	Method
	Location relative to markets
Conversion to secondary form	Feed mix
	Output mix (demand)
	Externalities (viz, environment)
	Method
Distribution	Demography
	Externalities (viz, remote siting)
Conversion to work-end function	Mechanical efficiency
	Capital stock energy component
	Input mix quality
	Performance requirements

Fig. 6. Path of energy use.

Examples of some end functions are environmental conditioning (heating, air conditioning, lighting), transportation, labor savings (as through appliances), recreation, and health maintenance (as through refrigeration, cooking, clothes washing and drying, and most domestic hot water uses). In intermediate processes, energy is used for physical or chemical separation or joining of materials, the gathering of materials (such as mining or harvesting), transportation of materials, and delivery of most of the final demand services listed above.

The external factors which affect the efficiency of each stage of the energy-use process (see Fig. 6) are, in some cases, included in published international comparisons. An example of such a factor is climate; a correction for degree-days in a heating comparison is typically made. Other external factors, usually omitted, include differing socially imposed externalities such as environmental requirements, quality differences such as time (the time savings of air travel versus bus transportation), and comfort and safety (small cars versus large cars).

In summary, quality differences exist among the processes and end functions, as in differences in refinery or steel mill product mix, or in the convenience differences between automobiles and mass transit. It is only after these quality differences have been recognized, and their contribution to energy demand analyzed, that comparisons of this sort can indicate areas likely to yield significant energy savings.

An example of such a comparison is shown in Fig. 7.[15] This comparison, between a gasoline-powered and electric-powered car, assumes vehicles of identical weight, presumed to offer equivalent utility. Comparisons between total systems such as these do indicate a measure of the resource consumption for a specific end function.

This example (Fig. 7) illustrates the second path to conservation as indicated on Fig. 8 — a process change. In this same vein, a conversion to diesel would

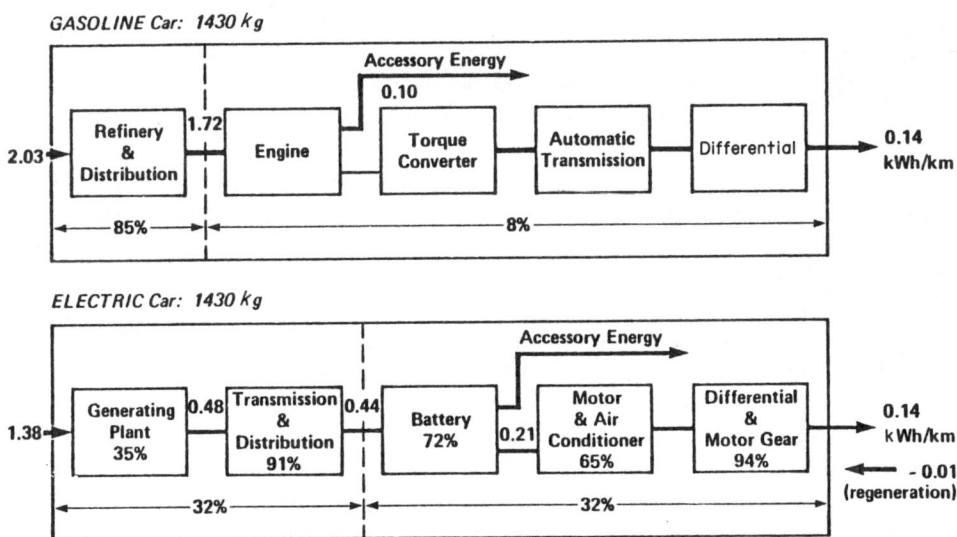

Fig. 7. System efficiency. Gas vs. electric car, federal driving cycle.

provide improvements in individual process efficiencies (as diesels are more efficient than gasoline engines and refineries use less energy in the production of diesel oil than of gasoline), as well as represent a process change.

The comparison and examination on a case-by-case basis of the primary energy-flow pattern in the system used to perform various end functions will serve to direct conservation efforts better than will the comparison of more aggregated data. Such an examination can indicate the relative importance of process efficiencies, sequences of processes, or quality changes in end functions.

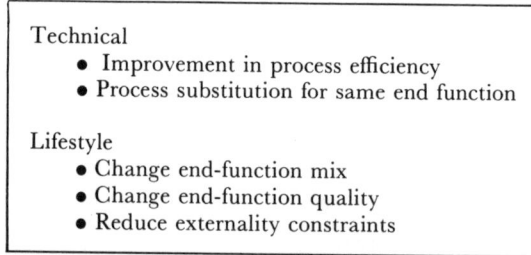

Technical
- Improvement in process efficiency
- Process substitution for same end function

Lifestyle
- Change end-function mix
- Change end-function quality
- Reduce externality constraints

Fig. 8. Technical and lifestyle conservation.

Quality of Energy Resources

One feature of interregional comparisons on a "per Btu" basis that limits their value to a national planning effort is the usual failure to discriminate between the relative performance of different fuels. While this is not a novel point (Boretsky has used weighting factors for various fuels), it needs to be addressed in these international comparisons.[5]

In the US today, we are developing methods to convert coal to clean liquid and gaseous fuels with "Btu" efficiencies of between 50% and 80%. The large-scale implementation of these systems would obviously increase gross energy consumption relative to the Btu content of the oil and gas, but given the large coal resources of the US such systems may be acceptable. Currently, coal and uranium are usable on a large scale only after conversion to electricity, and the necessary heat rejection during conversion is frequently mentioned as proof of the "inherent inefficiency" of electricity. While it is clear that, on a per Btu basis, the use of electricity for space heating requires more primary energy than the direct consumption of natural gas, it is not clear that the latter method is preferable if other than Btu comparisons are made. Figure 9 evaluates a number of heating strategies by the amount of oil and gas consumed.[14] In this figure, the oil and gas consumption for electric heating is based on their fraction of the total primary fuel used for electricity nationally. While this measure is too simple for general application, it does point out the value of alternative viewpoints as to what should be conserved, as well as how much. For example, if the objective is reduction in oil and gas use, and coal and uranium are unrestricted, electricity for space heating would be chosen.

The Btu measure of fuel value is appropriate for processes which lead to low-temperature heat, which represents the lowest valued use of fuel. Comparisons of primary energy use for other types of conversion, such as from potential energy to mechanical motion, high-temperature heat, or light are poorly served by this measure.

Even for applications with heat as an end function, this simple calorimetric comparison ignores the overall sequence efficiency of use for most applications. As an example, the energy potential of a high mountain lake can be converted to electricity with 90% efficiency, delivered with 91% efficiency, and used to generate heat with 100% efficiency (or more if a heat pump is used). This 82% system efficiency is substantially higher than the efficiencies found in any fossil-fuel system. As a result, most energy analysts have adopted a quality multiplier for hydroelectric energy to put it on the same basis as fossil-fuel electricity. While such a multiplier seems appropriate, it seems that other multipliers for other fuels would be equally appropriate.

The concept of weighting factors for different fuels leads immediately to inconsistencies. As an example, the energy value of oil relative to coal for electricity generation is roughly equivalent on a Btu basis, because the system efficiencies are about the same. The system efficiency of oil is much higher than that of coal if a comparison of transportation applications (diesel vs. steam) is made, and its use should be weighted accordingly. While it is theoretically possible to weight the value of each fuel by a factor which reflects the aggregate of end-use functions to which it is applied, such a factor would be inappropriate for specific end use comparisons.

The issue of resource quality is handled quite nicely by market economics when given a chance. It seems likely that the aggregate weighting functions for various fuels could be inferred from their free market prices. If this is true, then on a basis

Residential Scarce-Fuel Use (10^{12} Btu/yr)

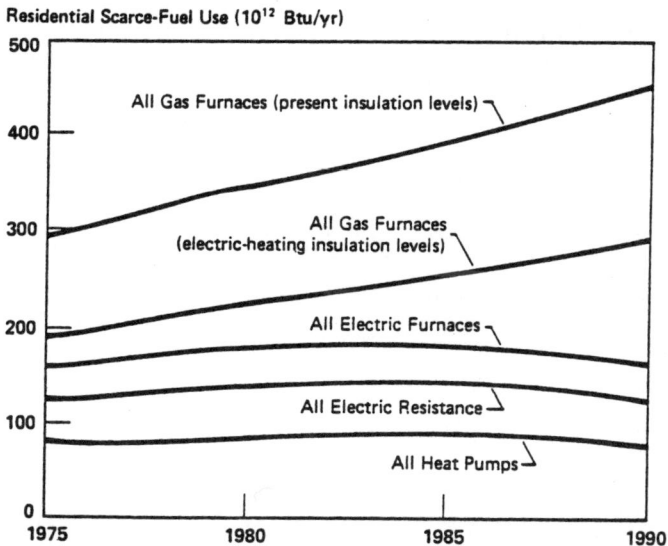

Fig. 9. Oil and gas consumption for future heating scenarios.

of social value per Btu, hydroelectric potential is more valuable than natural gas, gas more valuable than oil, oil more than coal, coal more than uranium, and uranium more valuable than sunlight. A wise conservation policy is one which recognizes these quality differences and does not measure success simply by the number of Btu's saved.

Conclusions

In the introduction, a number of international comparison studies were cited. The question, of course, is what are the implications of these studies? Do they provide us with guidelines, or alternative ways of using energy? Not in their present state of analysis. What they have shown is that for some comparable total bundle of goods and services energy used in a number of European countries is less than in the US. Part of this is due to the inter-industry mix, part to lifestyles, differences in prices of factor inputs, the prices of final products, incomes, climate, geography, natural-resource base, technology, and so on. The list is, indeed, formidable. But this list does suggest the complexities involved when trying to compare energy use in different societies. In other words, GDP/E ratios can be expected to be different in each country, reflecting the most efficient way to do things for that society. How they do things, how they combine their resources, how they use their energy does not mean that their way is the way we should follow. Is it not possible, should the US emulate any country, or combination of countries, that the mis-allocations of our resources could be large? Rather than emulate any country, market forces (prices) will do a lot towards adjusting and reallocating energy to its best use with conservation taking place in the process. The US economic system, its society, is capable of making change, adjusting to the new (higher) prices of energy. History has shown us again and again of this flexible capability.

To sum up: the problem of comparing countries and their respective energy uses is a complex one, and simple GDP/E ratio comparisons can lead to the wrong conclusions. Let us avoid that. Certainly, such ratios do not measure efficiency. Hopefully, our State comparisons will provide some insights into the complexity of the issues involved and the reasons for differences in energy use.

References

1. E. Berndt and D. Wood, An economic interpretation of the energy-GNP ratio, in M. Macrakis (ed.), *Energy: Demand, Conservation and Institutional Problems*, Chap. 3, Cambridge, MIT Press, 1974.
2. E. Berndt and D. Wood, Technology, prices, and the derived demand for energy, *Review of Economic Statistics*, Aug. 1975.
3. Ben-Chieh Liu, *The Quality of Life in the United States, 1970* (Midwest Research Institute, Kansas City, Mo., 1975).
4. Ben-Chieh Liu, *Quality of Life in the U.S. Metropolitan Areas, 1970* (Midwest Research Institute, Kansas City, Mo., 1975).
5. M. Boretsky et al., "Potentialities and Limitations for Conservation in the United States Apparent in Differential Uses of Energy Abroad" (Office of Policy Development, Office of the Secretary, Washington, DC, Revised Draft, 1975).

6. E. Cook, *Man, Energy, Society*, San Francisco, W.H. Freeman & Co., 1976.
7. J. Darmstadter *et al.*, *Energy in the World Economy*, Resources for the Future, Baltimore, The Johns Hopkins Univ. Press, 1971.
8. J. Darmstadter, J. Dunkerley and J. Alterman, *How Industrial Societies Use Energy: A Comparative Analysis* (Resources for the Future, Washington, DC, Preliminary Report, 1977).
9. R. Goen and R. White, *Comparison of Energy Consumption Between West Germany and the United States* (Stanford Research Institute, Palo Alto, CA, 1975).
10. J. Griffin and P. Gregory, An intercountry translog model of energy substitution responses, *The American Economic Review*, Dec. 1976.
11. J. Kendrick and C. Jaycox, The concept and estimation of gross state product, *The Southern Economic Journal*, (Oct. 1965, Vol. 32).
12. I. Kravis *et al.*, *A System of International Comparisons of Gross Product and Purchasing Power*, Baltimore, The Johns Hopkins Univ. Press, 1975.
13. B. Netschert, *Fuels for the Electric Utility Industry, 1971-1985* (National Economic Research Associates, Washington, DC, 1972).
14. J. Oplinger, Electric heating can save scarce fuels, *Electrical World*, 15 Oct. 1975.
15. J. Salihi, Kilowatthours vs. liters, *IEEE Spectrum*, Mar. 1975.
16. L. Scheer, "The quality of life: a try at a European comparison, in EPA, *The Quality of Life Concept: a Potential New Tool for Decision-Makers*, Washington, DC, US Government Printing Office, 1973.
17. L. Schipper and A. Lichtenberg, Efficient energy use and well-being: the Swedish example, *Science*, Dec. 1976.
18. S. Schurr *et al.*, *Energy in the American Economy, 1850–1975: An Economic Study of Its History and Prospects*, Baltimore, The Johns Hopkins Univ. Press, 1960.
19. C. Starr and S. Field, *Energy Use Proficiency: The Validity of Interregional Comparisons* (Electric Power Research Institute, Palo Alto, CA, 1977).
20. G. Stigler, *The Theory of Price*, New York, The Macmillan Co., 1952.
21. University of California, Lawrence Berkeley Laboratory, *A Linear Economic Model of Fuel and Energy Use in the United States* (A report to EPRI — ES-115, Dec. 1975).
22. Will energy conservation throttle economic growth? *Business Week*, (25 April 1977), pp. 66-80.

9

*Energy System Options**

Introduction

It is a frequent postulate that the availability and cost of energy are the greatest long-term problems faced by the industrialized countries of the world. Less often mentioned is the influence of energy supply on development in the nonindustrial countries. Energy costs and availability certainly will be key factors in the future of these regions.

The possible and preferred strategies for global energy development will be substantially determined by regional planning. Regionally, local benefits and penalties are the significant issues — the potential global problems from energy use (such as global weather changes from CO_2 production) are rarely given substantial weight in the decision process. The lack of institutional arrangements for dealing with global consequences of regional actions will be a growing problem as more areas of interaction are identified, and as the scale of worldwide energy development increases.

Within any regional planning unit the goal is the optimal use of resources integrated over time. The resources to be utilized are land, labor (both quantity and quality must be considered), capital as well as the existing capital stock, and the natural resources of materials including fuels. Every region of the world has always been limited in one or more of these resources.

Primitive food-gathering societies were limited by the natural rate of food production — but domestic agriculture removed this limitation. In the United States 100 years ago a fourth of all arable land was used for feed for draft animals — at that time extrapolation of agricultural trends a century ahead would have been extremely pessimistic. A hundred years ago, the solution to such resource limits was geographic expansion and technological development. The expansion to new lands has been largely exploited (unless we consider the advocates of space colonization seriously), but all evidence indicates that the technology frontier has no limits.[1] Although long-range future projections based on large changes in technical capabilities have a "science fiction" ring to them, our recent history shows that our ability to predict the direction of technical change is as yet poorly developed. Yet, uncertain as future scenarios are, the hazards of assuming an unchanging technological base are perhaps greater: consider this quote from 1922.[2]

*Presented at the 10th World Energy Conference, Istanbul, Turkey, 19–23 September 1977.

133

"According to Eugene Davenport, Dean of the Illinois College of Agriculture, the greatest future need of American agriculture is a fundamental
national policy. One of our leading statisticians estimates that a century
from now our population will amount to more than 225,000,000 people. The
prophecy is startling because it suggests possible hunger and even famine as
our future. At present, with less than half these numbers, our food production is only about equal to our domestic consumption. Unless we institute
very revolutionary practices to enhance production, we may look for it to fall
behind. Present indications are that we are rapidly slipping into the class of
food-importing nations. This means that unless we are able to reverse the
tide, we must readjust our social, economic and industrial organizations to
accord with this new condition."

The outcome did not fulfill the prophecy — the United States reached the
stated population in roughly 50 years, not 100, yet is the leading exporter of food
in the world. The missing ingredient in this projection and many like it was the
recognition of the important technological changes which were already pending
at the time of the projection.

In discussing energy futures it is important to include not only technical options
but also the adaptability of social and political systems to resource limitations,
and the technological choice stimulated by these limitations. Certainly projections of the US social structure made before the automobile would seem ridicul-
ous in hindsight, yet the enormous changes in demography and social structure
due to the automobile occurred principally within the space of 50 years, the first
half of this century. The time frame in which the integration of new energy
systems occurs is also roughly 50 years, and therefore planning horizons of this
duration should be the norm in studies of new energy options. The relative times
and costs for the phases of energy system growth are shown in Fig. 1. Only the cost
estimates on this figure have been changed from the original in 1971.[3]

It is precisely the social, political, and technological adaptability of societies
that makes future energy predictions difficult. Although it is tempting to focus on
shortages of fuels and raw materials as setting limits in an eventually steady-state
world,[4] the more pressing problem in underdeveloped areas is the transient
dynamics of maintaining long-term economic growth. In most cases the short-
term limiting resource is capital.[5] Within the framework of total resource utilization and development, it is the issue of how scarce capital should best be applied in
the capital intensive energy area that this paper addresses.

Part I. Traditional Energy System Assessment

The energy system characteristics traditionally analyzed in the industrialized
countries provide a basis for a regional assessment. First among these is cost. The
capital costs per unit of energy-supply capacity will most likely be the single most
important factor, particularly in countries with very limited capital. Operating
costs, involving fuel and system reliability, do trade-off with capital costs, depending on capital availability. The long-term social costs arising from environmental
impacts of energy systems should be considered in terms of the present value of
future risks and benefits, but these are more difficult to evaluate and assess.

Control	Individual selection	Societal selection	Economic feasibility	Technical feasibility	Natural limitations
Implementation time (years)	1	10	10-100		100-1,000
Costs involved (dollars)	10^2-10^4	10^6-10^9	10^9-10^{11}		

Individual selection

OPTIONAL USES
Comfort (heating, air conditioning)
Entertainment
Communication
Home
Transportation
Labor aid
CRITERIA
Relative costs
Personal safety
Quality of life
Intangible and subjective biases

Societal selection

DEVICE UTILIZATION
Central station versus local power plant
Type of conversion method
Distribution alternatives
RESOURCE DEVELOPMENT
Coal
Oil and natural gas
Nuclear
Shale oil
Coal gasification
Fusion
Solar
SITING CHOICES
Origin of fuel
Close to user
Consider aesthetics
Land-use alternatives
Waste disposal
Environmental deterioration
REGULATION AND CONTROL
Legislation
Regulations
Standards

Economic feasibility / Technical feasibility

SPECULATIVE RESOURCES
Solar power
Fusion
Biological photosynthesis
Fuel cells, MHD, direct conversion
ALTERNATIVE FUELS
Alcohol
Liquid hydrogen
Ammonia
ENVIRONMENTAL EFFECTS
Recycle wastes
Waste storage (radioactive)
Underground distribution
Safety

Natural limitations

RESOURCES
Finite fossil-fuel reserves
Uranium usage depends on breeder
CONTINUOUS SOURCES
Limited except for solar
ENVIRONMENTAL EFFECTS
Thermodynamic limit on conversion efficiency
Regional climatic effects
CO_2 production inevitable from fossil fuels

Fig. 1.   Controlling factors that enter into large-range planning.

The initial factor in all these issues of energy system planning for underdeveloped areas is capital availability. Besides the normal competition for economic output between consumption goods and capital formation, the competing industrial and public uses for available capital will limit that available for energy-system expansion.

Capital Availability Calculation

A rough estimate of the potential rate of installation of an energy system can be inferred from a simple capital availability calculation, except for those rare countries in which capital is not a serious constraint to development (e.g. the OPEC countries). The fraction of economic output used for gross capital formation tends to increase slowly with increasing per capita Gross Domestic Product (GDP). Data from reference 6 suggests that typical capital formation rates are 5–15% of GDP in the poorest countries (those with GDP per capita of less than $100 US) to 15–300% for the industrialized nations. A sizeable portion of this capital represents depreciation reserves to cover the ultimate replacement of existing stock, and as a result represents capital resources that are not available for expanded facilities.

If 20% of gross domestic product is taken as an optimistic fraction for capital projects, and of that, 15% goes for capital purchases of energy systems (relatively representative of the pattern in the US), then 3% of gross domestic product is invested in energy system capital stock each year. Under these conditions, the derivation of future energy system capital stock is quite straightforward (see Appendix I) and gives

$$Sc(t) = S_{co} e^{-(I_d + I_p)t} + \frac{.03 \text{ GNP}_{co}}{(I_d+I_g)} \left[e^{(I_g-I_p)t} - e^{-(I_d+I_p)t} \right]$$

in which

$Sc(t)$ = energy system capital stock per capita at time t,
S_{c0} = energy system capital stock per capita at $t = 0$,
GNP_{c0} = GNP per capita at $t = 0$,
I_d = depreciation rate,
I_g = GNP growth rate,
I_p = population growth rate.

The results of this equation, which assumes purely exponential growth of population and GNP, are shown in Fig. 2. An assumption implicit in the model used for this figure is that economic output will rise in direct proportion to capital stock (hence the purely exponential growth of both GDP and capital stock). For the industrialized countries, this assumption is well supported by the data.[7] Calculation of capital-output ratios (the ratio of capital stock to GDP) for the developed countries is remarkably constant.

For the under-developed areas, generalizations cannot be made; the capital-output ratio can be either higher or lower than that of industrialized nations.[8] The capital-output ratios of the economies of under-developed regions are subject to two effects not applicable to developed economies. First, the initial units of capital investments, if well chosen, can yield greater returns in under-developed than in developed countries largely because the initial units of capital invested in any economy are inherently more productive than subsequent investments.[8] The second factor in under-developed areas is the need for social overhead capital such as roads, communication systems, and energy systems. These investments are referred to in the economic literature[5,7-10] as indivisible or "lumpy", in that there is a certain threshold cost for these systems that is independent of their use and thus their expansion and use is discontinuous. Yet these systems are needed if other sectors of the economy are to develop. These two effects tend to counteract each other; the initial high marginal return on investment tends to push the capital-output ratio down, and the need for social overhead capital tends to raise the ratio. A consensus of the literature[5,7,9] suggests that the capital-output ratios in under-developed areas will be slightly higher than in industrialized countries.

Determining optimal methods of resource allocation is extremely complex, yet the issue of energy system capital allocations can only be considered within the context of the total problem of development. Assessment of energy options for the under-developed countries is more closely coupled with total resource development decisions in the under-developed regions than in the developed areas. For

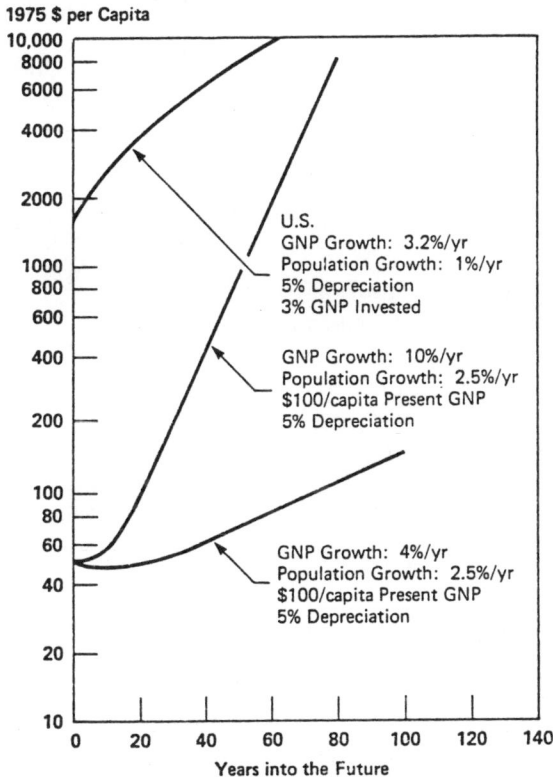

Fig. 2. Growth of energy system capital stock.

the under-developed countries, the social and technological changes that occur with economic development may determine the dynamic process of energy-system growth in such countries.

Part II. Energy Option Assessment in an
Economic Development Framework

Energy will be required for economic development. In a region with very limited energy resources, or with limited capital to utilize its resources, care must be taken to apply the available energy to the most productive functions. A distinction can be drawn between the energy used to meet direct consumer demand and that used indirectly as an input into a production process. While the final direct energy produces substantial amenity benefits to the user, comfort (as from space heating) or savings in domestic labor, the focus in under-developed countries is likely to be on the other indirect uses — energy for production.

The factors which typically contribute to increased productivity — capital stock, division of labor, distance to the markets for the goods produced — all recommend that development efforts be directed to cities. This strategy carries the advantage of best utilizing the existing social overhead capital of roads and energy systems. Concentrations of capital stock will maximize the return on new social overhead capital, and additionally such concentration permits economies of scale in all of these areas (discussed in more detail later).

The type of end use application of energy will impose a constraint upon the selection of an option. Much has been written about simple small scale energy systems for under-developed areas (see, for example, reference 11), and for rural areas these small scale systems appear quite attractive for a number of reasons — much smaller initial investments are required, less specialized training is needed for construction and maintenance, and the cost of energy transportation is reduced or eliminated. However, while these small systems will improve rural amenities, they make little contribution to GDP. The economic incentive to seek development first in urban areas, and to concentrate on energy for production, leads to a type of energy demand that is much different than that of the rural areas.

In the urban areas, the centralized systems have two substantial advantages. The first advantage is that of lower cost. Based upon the costs of energy systems in place (as contrasted with theoretical analysis), energy delivered by wire or pipeline from a central facility is considerably cheaper than energy from small systems, for urban and industrial centers. A recent editorial in *Science*[12] which recommends rapid development and utilization of small solar systems for rural areas indicates what some of the typical costs are:

> "The cost of utility power in the United States averages 3 to 10 cents per kilowatt hour. It runs as high as 45 cents/kWh in urban areas of developing countries. In rural areas, however, power is available only from diesel generator sets at $1/kWh or more, or from primary batteries at about $12/kWh. Complete solar-thermal power systems costing about $4 per peak watt and capable of providing electricity at less than $4/kWh are already available. Photovoltaic systems costing $1 to $2 per peak watt are expected

by 1980. The OTA report* says that solar devices capable of providing on-site power at much less than $1/kWh could be produced in the next few years."

The cost of urban electricity can, with sufficient load density, be significantly cheaper than "45 cents/kWh". For any rural production process, an energy cost of $1/kWh is an enormous competitive disadvantage. (And these cost estimates did not include energy storage costs.) At this cost, a simple light bulb, run for 10 hours a day, uses over $300 of electricity a year. In many underdeveloped and developing countries, a GNP per capita of less than $300 is common. It would take all of the economic output of the typical laborer just to pay for this energy. This must be viewed as an investment with a very low return relative to its cost. Much more worthwhile investments can be made in urban energy systems with the same amount of capital.

The second advantage of the centralized energy systems is due to the form of energy produced. For most production processes, the demand is for mechanical motion (as from motors), high-temperature heat, and light. These requirements can be satisfied by electricity, but not by low-temperature energy sources such as solar space heating. For space-heating requirements in urban and industrial areas, utilization of waste heat and cogeneration appear to be more attractive sources than do the small systems. The one significant exception is that of energy for irrigation, which represents a highly disaggregated load. Improvements in agricultural yields can, in addition to the primary benefit of reducing hunger, contribute favorably to national economics by reducing food-import requirements or generating income through export,[8] providing means are available for the transport of agricultural products. For irrigation, small diesel or wind-driven pumps may represent favorable investments.

Costs of the Options

In considering the choice of energy options, the only systems for which the costs are known with certainty are those that are commercially available. The costs of the new technological energy sources are quite uncertain; some options are expected to become much less expensive due to research and development (solar photovoltaic, for example) while other options show less promise for reducing present cost estimates (ocean thermal gradients and wind power are examples). With the uncertainty due to changing technical capabilities, as well as uncertainties due to future fuel prices, estimates of future energy costs and capital requirements are very tentative. The capital costs of currently available systems are displayed in Fig. 3.[13] These costs are for the United States, in 1976 dollars. Figure 4 indicates the projected cost targets of electrical capacity using coal and coal derived fuels using technologies that are currently under development.[14] The electricity costs of these systems, in 1976 dollars, are indicated in Fig. 5. These cost estimates are calculated using 1975 fuel prices.

---

*Application of Solar Technology to Today's Energy Needs, Office of Technology Assessment, Washington, DC, June 1977.

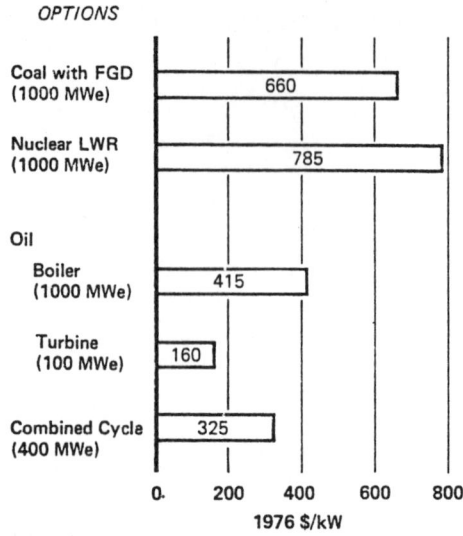

Fig. 3. Capital costs: current electric power options.

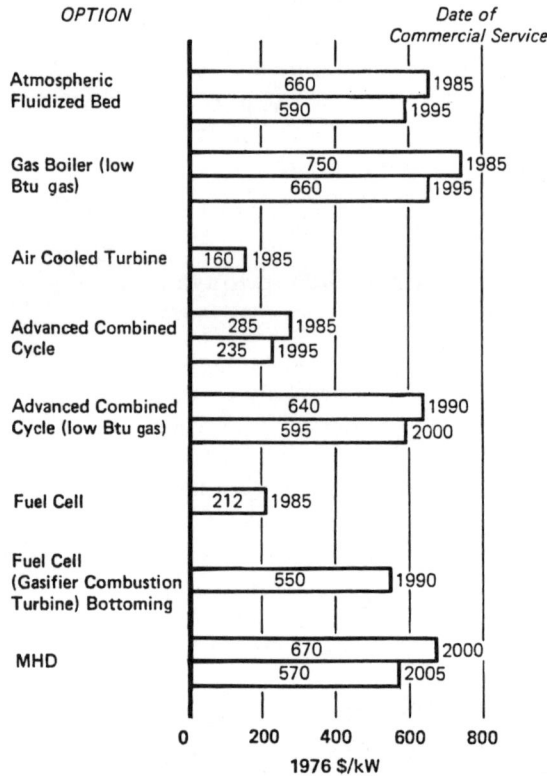

Fig. 4. Capital costs: fossil fuel electric options under development.

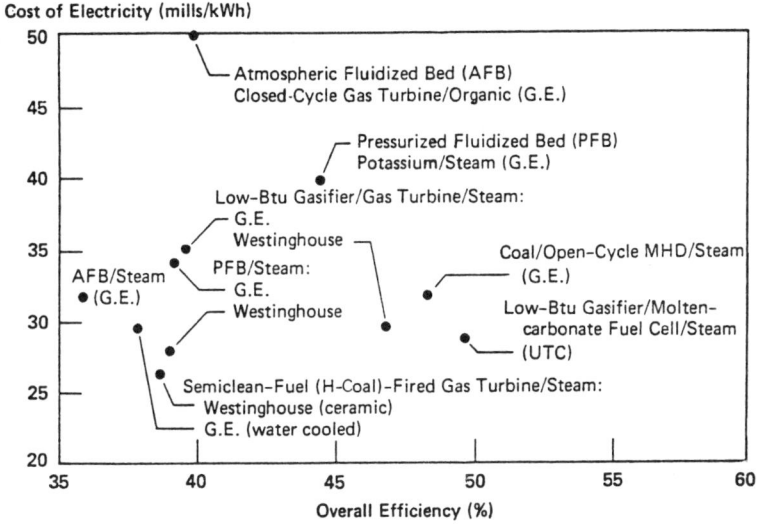

Fig. 5. Advanced fossil-fueled electric power generator: projected cost.

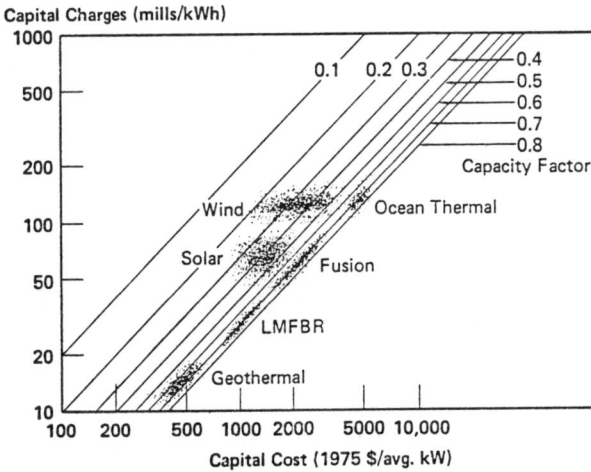

Fig. 6. Capital costs and capital charges for new electric power generation options.

Energy costs from electric power options under development which do not use fossil fuels will depend almost entirely upon the capital cost and capacity factor, as indicated in Fig. 6. Hydroelectric was not included because the costs are quite site specific; photovoltaic was excluded because present high costs are expected to drop considerably due to new research results, but the uncertainties are simply too great. Based upon the present level of technology, Fig. 7 gives a classification of all the significant energy sources now in use or under technical development.

Low	Medium	High
DIRECT ENERGY	DIRECT ENERGY	DIRECT ENERGY
Coal	Biomass conversion	Electrolytic hydrogen
Oil	Coal liquefaction	
Gas	Wind for pumping	ELECTRIC CONVERSION
Farm waste		Wind electric
Solar heat	ELECTRIC CONVERSION	Ocean thermal
Coal gasification	Solar thermal electric	Satellite solar
	Fusion	Photovoltaic solar
ELECTRIC CONVERSION	Diesel	
Coal	Biomass	
Fission converter	Oil	
Fission breeder	Gas	
Hydroelectric		
Geothermal		
Municipal waste		

Fig. 7. Relative cost of energy production (excluding transportation).

Option Assessment — Infrastructure

A key factor in determining energy-system capital requirements is the cost of developing the infrastructure required to utilize a particular energy technology. It is this factor that contributes substantially to the popularity of oil — an oil-based system requires a very low investment in systems to transport energy. If the oil is imported, a refinery is not even needed, as various oil derivatives can be imported directly; the fuel is easily stored, and the level of technical sophistication required to operate on an oil-fueled economy is much lower than that required for a nuclear or coal-based economy.

As an example of the infrastructure required for a coal-energy system, a recent report[15] estimated that, in order to double US coal production (600 million tons per year), the following actions would be needed:

Develop 140 new 2-MTPY eastern underground mines.
Develop 30 new 2-MTPY eastern surface mines.
Develop 100 new 5-MTPY western surface mines.
Recruit and train 80,000 new eastern coal miners.
Recruit and train 45,000 new western coal miners.
Manufacture 140 new 100-cubic-yard shovels and draglines.
Manufacture 2400 continuous mining machines.

These items are in addition to the tremendous capital requirements for coal-fired electric power plants and the transmission and distribution system needed to deliver the energy, or, alternatively, the coal gasification plants with a pipeline network. It is probable that for many countries the number of engineers could be a limitation, so the cost of creating engineering schools or of sending students to other countries must also be included.

For a nuclear system, the material flow is significantly lower than for coal, but the number of areas in which technical specialization is required is greater. A number of developing countries have recognized this problem and established local training facilities to provide the necessary manpower. India has already trained more than 3000 nuclear engineers, and ambitious plans are underway in Brazil, Iran, and Pakistan.[16]

There can be capital costs that go beyond these issues of industrial infrastructure. When the Tennessee Valley Authority began operations in the United States, it was found that many people in the TVA service area could not afford many of the basic energy-using devices. As a result, TVA issued guarantees on extremely low interest loans to stimulate purchase of such items as irrigation pumps and refrigerators. Thus capital was needed for the end-use devices as well, increasing overall requirements.

Effects of Energy System Size

From the preceding discussion, it is clear that the capital required for a complete energy system greatly exceeds the traditional power plant costs. The cost of these systems will also depend on scale factors both in energy distribution and, in the case of electric power plants, unit size.

Transportation of energy can be divided into large-scale bulk-fuel shipments and local distribution. As Fig. 8 indicates,[17] large-scale transportation of fossil

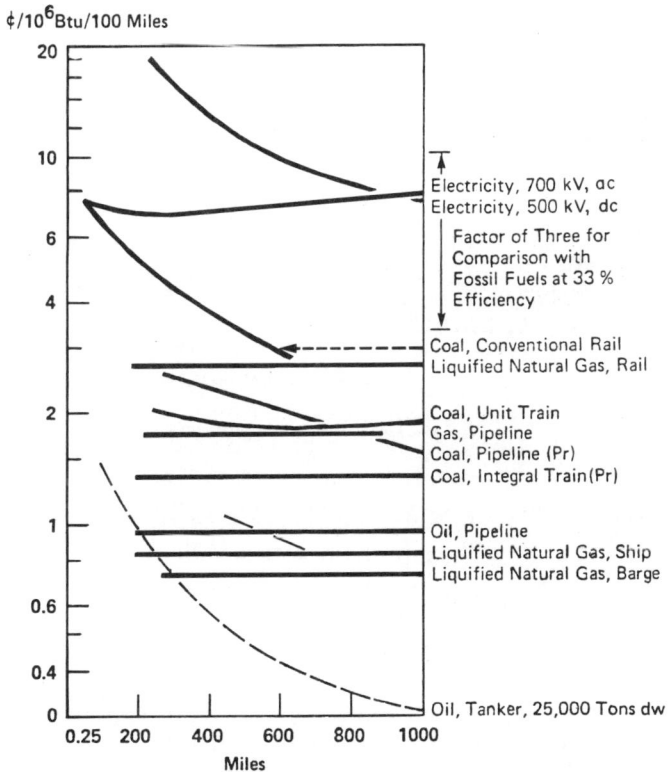

Fig. 8. Transportation costs.

Capital Stock (rate base items) Per Customer

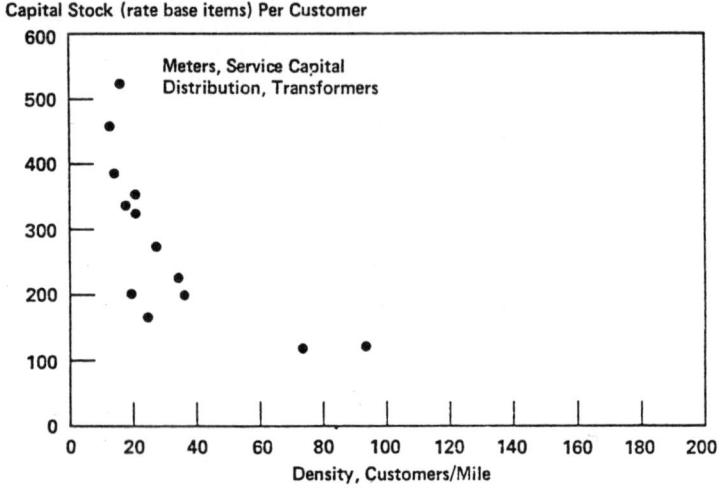

Fig. 9. Distribution capital stock vs. customer density.

Capital Stock/Customer

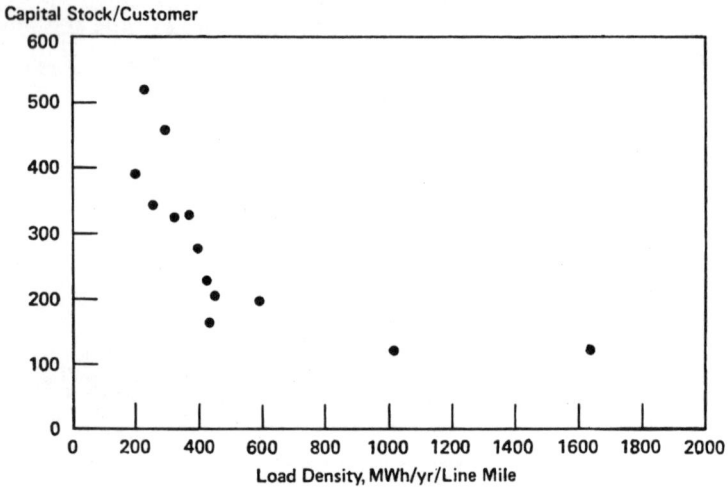

Fig. 10. Distribution capital stock vs. load density.

fuels over long distances is inexpensive relative to the value of the fuel (the transportation cost is several cents per million Btu's; the world price of oil is several dollars per million Btu's). The more important issue for developing countries is the local distribution of energy. Some minimum capital investment is needed to install a system for energy distribution to each household. For fixed systems such as those associated with electrical systems or gas pipeline systems, the capital costs of these systems is relatively insensitive to the quantity of energy delivered, but rather depends upon the density of energy consumers. For the rural areas of the underdeveloped and developing nations, this capital cost will be very high. For example, rural electrification in the US was a very expensive public works project in the 1930s; it was heavily subsidized by the rest of the US economy to meet the sociological objective of economic growth of the rural areas. These rural areas show distribution system capital costs of about $1000 per customer. On purely economic grounds, at the time it was built, rural electrification could not be justified.

Capital costs for electrical distribution are quite sensitive to population and energy use density, as indicated in Figs. 9 and 10.[18] In these figures, the capital stock numbers refer to acquisition costs; the replacement costs would be much higher.

In rural areas with low energy density a qualitative description of energy distribution costs is of great importance (see Fig. 11). For these areas, transportation costs will be a highly significant factor in both the total production costs and the total capital requirements.

Low	Medium	High
Wind	Oil	All central station
Solar heat	Kerosene	electric
Farm waste	Synthetic liquid fuels	All pipeline gases
(all local use)	Propane	
	Diesel electric	

Fig. 11. Energy transportation costs for low energy density regions.

For the urban centers in these countries a different situation exists: the energy densities are quite high (as is the population density), so transportation is much less significant an issue. Additionally, energy use in urban areas is likely to be a more effective producer of economic output. The urban areas have more infrastructure with a resultant spread of specific labor skills and lower transportation costs for materials. For rural areas in underdeveloped countries, the capital required for transportation of central station power is so large as to overwhelm the choice of the energy conversion system.

In urban areas, generating equipment scale factors will also make a significant impact on capital requirements, as shown in Fig. 12.[19] From this reference, the unit cost sensitivities can be computed in the form

$$\text{cost}/kW = C_0 \, (KW_0/KW)^\alpha$$

in which C_0 is the cost per kW of the reference plant of size KW_0, KW represents the design size, and α is a scale parameter. For a light-water reactor $\alpha = .31$ [19] which means that a doubling of plant size only increases plant cost by about 60%. For fossil-fueled plants the scale effect is smaller; α is typically between .15[13] and .25.[19]

The implication is clear — for those areas in which capital or demand limits construction to small plants, the energy costs will be higher. Combining the distribution and production costs of energy, it seems likely that energy will be

Unit Capital Cost [$/kW(e)]

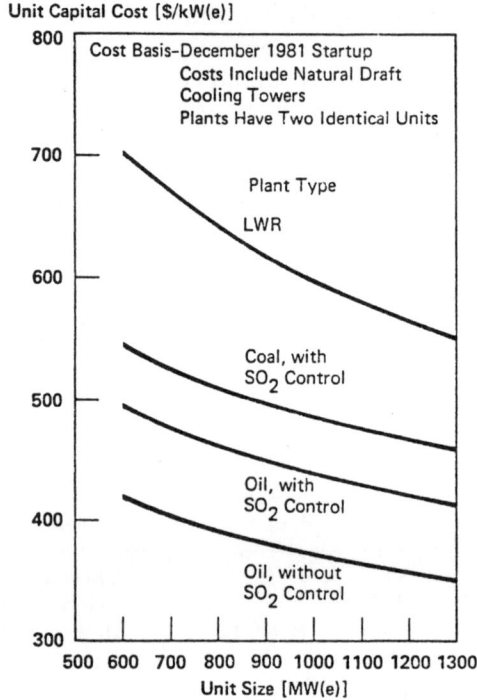

Fig. 12. Unit capital costs of power plants as a function of unit size.

most expensive in the least developed areas of the world, and in their rural areas. As a result, economic production activities which require a significant energy input will be at a competitive disadvantage in these areas.

Small-scale Energy Systems

The small-scale energy systems of which solar heating is an example can, for the reasons relating to energy distribution, be cheaper in areas with low density of energy use, yet the capital requirements for these systems are high, and the energy provided is more likely to contribute to comfort than productivity growth. In these rural areas, the use of farm waste is traditional, and there are technologies available at low cost which permit more efficient use of such waste (biogas digesters, for example). It should be recognized that large-scale biomass conversion to energy does not generally result in a net energy output, due to all the energy requirements for growth and harvesting. Some exceptions, such as sugarcane, can be found to be net energy producers. However, where biomass waste is already collected and available it certainly can be used locally.[20] Large-scale biomass projects are likely to compound the problems of global food supply. Energy plantations would compete directly for agricultural land. The following quote indicates a key issue of land use:[21]

"Two-thirds of the world's agricultural land is in permanent pasture, range, and meadow, of which 60 percent is not suitable for cultivation. These lands, where not claimed for other uses, are exploited best by large ruminant livestock, primarily beef and dairy cattle, which now produce almost half of the world's meat products and most of its milk products. Animals constitute an important stockpile of food and capital for the developing countries. The total caloric worth of worldwide animal stocks in 1974 was 50 per cent more than that of grain stocks, and the animal stocks were more evenly distributed."

Diesel power generators can have special applicability. For high energy density areas, the extremely high fuel costs and maintenance costs make diesel noncompetitive with centralized electric technologies. In areas where diesels are the primary source of electricity, as in a number of Alaskan villages, the costs range from 14¢ to 48¢ per kWh,[22] much higher than typical US power costs of 4–5¢ per kWh. Yet for these villages this is the cheapest electricity source available. The capital costs of such a system are quite reasonable — diesel generators are available in the US that retail for about $100/kW (in 1977 dollars) for 100 kWe output. Little additional capital is needed for ancillary equipment — fuel tanks and fuel-delivery trucks are not particularly expensive. While such a system can provide amenities, because of its very high fuel and maintenance costs, this type of power system will not lead to light industry that can compete with the urban industries using much cheaper central station power; and capital outlays of this type are made at the cost of more productive investments.

The most important exceptions to the high energy cost in rural areas arise from indigenous resources. Large hydroelectric facilities coupled with energy-intensive industry such as aluminum production can provide substantial economic benefits. Dry steam geothermal can also be used as an inexpensive regional power source. For these cases of a specific indigenous resource, the entire schedule of energy production and transportation costs is altered, and regional development plans centered around those resources can be extremely successful.

Implications of Scale Effects

Of all the social overhead items that require large initial investments, energy systems are, in the economists' jargon, the "lumpiest". A large power plant, with the related transmission and distribution network, frequently represents an investment in excess of a billion dollars. Most underdeveloped countries cannot allocate capital in this amount to any single project. Two alternatives present themselves. The first is to create a flow of capital to these countries from the industrial and resource-rich areas of the world. In many under-developed countries aid of this sort is the largest source of capital. While the political issues raised by this method of accumulating capital are beyond the scope of this paper, it represents an important avenue that should not be overlooked.

The second approach is based upon self-generation of capital, and generally produces slower results. A country which achieves positive economic growth generates increasing levels of capital, but the initial amounts accumulated are small. This in turn implies that smaller units will be purchased initially, and the economies of scale mentioned earlier will not occur. Resultant energy costs will be higher, and productivity growth lower than if the efficiencies of large scale were available.

Using the assumption of capital availability of the earlier calculation (that 3% of GDP would be available for energy systems) the energy system capital can be easily calculated. Figure 13 illustrates the relative amounts of capital for countries with GDP per capita of less than $500 (1972 data[6]). As these capital levels represent annual accumulations, the actual net worth of a multi-plant energy system will be larger (because it represents several years accumulation).

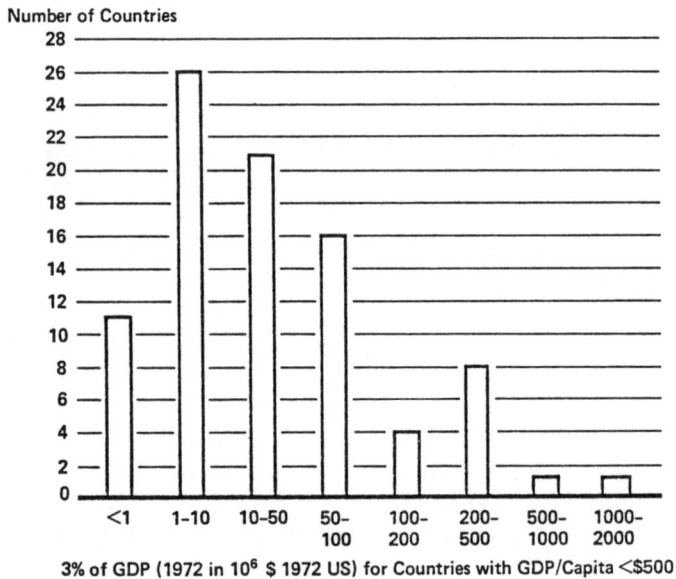

Fig. 13. Large-project capital capabilities, less-developed countries.

If economic growth is the overriding goal, the strategy suggested by this analysis would be to develop one area at a time. This would best utilize the efficiency of scale, and also serve to produce the infrastructure needed for productivity increases. Several important issues are raised by such a program.

Equitable distribution throughout a nation of the costs and benefits of the capital-intensive system might make such a policy politically difficult. The decision to pour capital into a single area would in many cases require the deferment of goals for other areas. Additionally, the risks of failure are magnified relative to a wider distribution of capital. As an example, several large power plants are cheaper than the equivalent capacity of smaller units, yet the impact and cost of a failure in a large plant may produce much more severe social consequences.

Industrialized and High Capital
Availability Regions

In view of the background presented on the relationship between energy systems and regional development, both the industrialized countries as well as the capital-rich nations can take actions which assist the under-developed regions to achieve their economic goals, while satisfying their own requirements. The spectrum of strategies is shown in Fig. 14.

The industrialized countries with capital are and will continue to be the technological leaders of the world, and are likely, for economic reasons, to reduce their resource consumption through the use of the high technology, capital-intensive substitutes which they can best apply. High technology development is a strategy that the industrial nations short on capital cannot afford. These countries will use coal if they have the resource, nuclear converter reactors, and oil and gas. The developing and under-developed countries with ample capital (OPEC countries, for example) have the greatest flexibility and widest range of opportunities. An approach to development like that taken in the Tennessee Valley of the United States in the 1930s is a likely choice for many of these countries. Under these conditions, energy-system development, coupled with programs for increasing labor productivity, can produce favorable long-term prospects.

| Economic type | Region type | Capital availability | |
		High	Low
INDUSTRIALIZED	All Areas	Responsible for technical leadership, has capital and infra-structure to utilize high technology sources, seeks to minimize resource competition with underdeveloped	Use lowest cost system that takes advantage of existing stock Deemphasize long-range energy R & D
UNDERDEVELOPED & DEVELOPED	Urban	Total city planning needed Advanced technologies	Energy density may justify central stations Development of industrial infrastructure is primary goal
	Rural	TVA type approach, smaller capital-intensive units, regional industrial centers established con-currently with agricultural capital expenditures	Improve quality of life with small systems—hyd-ro, farm waste, biomass, wind, solar

Fig. 14. Energy strategies.

In summary:

(1) Electrification of rural areas is a luxury that industrialized countries have achieved only recently — it is economically counter-productive for developing and under-developed countries.

(2) Advanced energy technology development is the responsibility of the high capital countries.

(3) Centralized energy sources and concepts will apply in all areas except rural areas of less developed countries.

(4) "Backyard energy sources" for rural areas can improve comfort and lifestyle but are unlikely to produce substantive productivity increases.

(5) Urbanization to increase energy density is the key to development — the cities are very cost effective due to economies of scale.

References

1. C. Starr and R. Rudman, Parameters of technological growth, *Science*, 182 (1973).
2. *Scientific American*, Dec. 1922.
3. C. Starr, Energy and power, *Scientific American*, p. 37, Sept. 1971.
4. A. M. Weinberg, *The Economics of Developing Countries*, Chaps. 6 and 7, Praeger, 1966.
5. H. Myint, *The Economics of Developing Countries*, Chaps. 6 and 7, Praeger, 1966.
6. *Yearbook of National Accounts Statistics 1974*, Vol. III, International Tables, United Nations, New York, 1975, St/ESA/STAT/SER.0/4/Add.2.
7. P. A. Samuelson, *Economics*, McGraw-Hill, 1976.
8. W. Letwin, in *Development and Society, The Dynamics of Economic Change*, David Novack and Robert Lebachman, eds.
9. G. M. Meier and R. E. Baldwin, *Economic Development, Theory, History, Policy*, John Wiley & Sons, Inc., 1959.
10. T. W. Swan, Economic growth and capital accumulation, in *Readings in the Modern Theory of Economic Growth*, Joseph E. Stiglitz and Hirofumi Uzawa, eds., MIT Press, 1969.
11. A. Makhijani, *Energy Policy for the Rural Third World*, International Institute for Environment and Development, Sept. 1976.
12. A. L. Hammond, An international partnership for solar power, *Science*, **197**, no. 4304, 12 Aug. 1977.
13. *Technical Assessment Guide*, EPRI, June 1977.
14. *Evaluation of Phase 2 Conceptual Designs and Implementation Assessment Resulting from the Energy Conversion Alternatives Study (ECAS)*, by National Aeronautics and Space Administration, Lewis Research Center, Report number NASA TM X-73515, April 1977.
15. *U.S. Energy Prospects: An Engineering Viewpoint*, National Academy of Engineering, Washington, DC, 1974.
16. M. Khan, Nuclear power and the developing countries, International Conference on Nuclear Power and Its Fuel Cycle, held by the IAEA in Salzburg, Austria, May 2-13, 1977, reported in *Nuclear News*, p. 79, July 1977.
17. *Energy Facts*, prepared for the Subcommittee on Energy of the Committee on Science and Astronautics, US House of Representatives, Ninety-Third Congress, First Session by the Science Policy Research Division, Congressional Research Service, Library of Congress, Serial H, taken from Technology Review, *Energy Technology to the Year 2000, A Special Symposium*, MIT, Cambridge, Mass., 02139, Oct./Nov. 1971, 48 pp.
18. Private communication, D. Speir, Pacific Gas and Electric Co., San Francisco; National Rural Electric Cooperative Association Paper 77-4, June 1977; and 1975 *Annual Statistical Report, Rural Electric Borrowers, Rural Electrification Administration Bulletin 1-1, Calendar Year Ended December 31*, 1975.

19. WASH 1345, *Power Plant Capital Costs, Current Trends and Sensitivity to Economic Parameters*, USAEC, Oct. 1974.
20. C. W. Lewis, Fuels from biomass — energy outlay versus energy returns: a critical appraisal, *Energy*, Vol. **2**, 241-248, 1977.
21. *World Food and Nutrition Study: The Potential Contribution of Research*, National Academy of Sciences, ISBN 0-309-02628-8, 1977 (from *News Report*, NAS, Aug. 1977).
22. Private communication, R. Lapp.

Appendix I

Calculation of Energy System Capital Stock

Notation

$S(t)$ capital stock at time t,
$S_c(t)$ per capita capital stock at time t,
$GNP(t)$ gross national product at time t,
$GNP_c(t)$ per capita GNP at time t,
$P(t)$ population at time t,
I_g GNP growth rate,
I_p population growth rate,
I_d depreciation rate

where

$$GNP(t) = GNP(0)\ e^{I_g t},$$
$$P(t) = P(0)\ e^{I_p t},$$

Initial values
The subscript 0 refers to initial values (as in $S_{c0} = S_c(0)$).

Assumptions
1. Exponential growth of GNP and population as described above.
2. Three percent of GNP invested in energy systems each year.
3. Depreciation of existing stock at I_d rate.

Derivation
The equation appropriate to the above assumptions is

$$\frac{dS(t)}{dt} = .03GNP(t) - I_d S(t). \tag{1}$$

(Capital stock is increased by the new investment, decreased by depreciation.)
Substituting for $GNP(t)$ and rearranging,

$$\frac{dS(t)}{dt} + I_d S(t) = .03GNP_0\ e^{I_g t}, \tag{2}$$

with

$$S(0) = S_0.$$

The unique solution to this simple first-order differential equation is

$$S(t) = S_0 \, e^{-I_d t} + \frac{.03 \mathrm{GNP}_0}{I_g + I_d} \left[e^{I_g t} - e^{-I_d t} \right]. \tag{3}$$

On a per capita basis

$$S_c(t) = \frac{S(t)}{P(t)} = \frac{S(t)}{P_0 \, e^{I_p t}} \tag{4}$$

$$= \frac{S_0}{P_0} e^{(-I_d - I_p)t} + \frac{.03 \mathrm{GNP}_0}{P_0(I_g + I_d)} \left[e^{(I_g - I_p)t} - e^{(-I_d - I_p)t} \right], \tag{5}$$

$$S_c(t) = S_{c0} \, e^{(-I_d - I_p)t} + \frac{.03 \mathrm{GNP}_{c0}}{I_g + I_d} \left[e^{(I_g - I_p)t} - e^{(-I_d - I_p)t} \right]. \tag{6}$$

10

Energy Planning – A Nation at Risk

At the turn of the century, if historical trends continue unabated, the United States will consume more than twice as much energy annually as it now does. The fuel resources needed for electricity generation would be more than four times our present annual usage. Clearly, if the US is to seek a path differing substantially from these trends, its chosen course should be guided by an assessment of the broad social, economic, and resource consequences of energy policy alternatives — rather than as an unintended and unforeseen happenstance of piecemeal project decisions arising from disparate economic, environmental, and political expediencies.

Although the United States has suffered only occasional short-term shortages of certain fuels, the economic impact in these cases has been quite measurable. Insufficient supplies could result either from a failure to build energy-producing and processing facilities such as electrical generation stations, oil refineries, or coal mines — or from another oil embargo. A dramatic rise in future energy prices could have a similar effect. Uncertain availability and cost of future energy supplies also create an unfavorable atmosphere for capital investment, and this inhibits our national economic vitality.

Those responsible for planning the future national energy supply have viewed it traditionally as an essential service for the health and survival of our economy. In contrast to this view, the classical economist treats energy just as any other economic variable rather than as something requiring special analysis. This implies that demand and supply are fully responsive to price, and the economic system can adjust fully to changing price by freely substituting other things for energy. This simplistic approach ignores the possibility that the role of energy in the economy may be different than most other goods, for which there may be a large array of substitutes. If energy is more like some basic necessity such as food, then shocks to the energy supply may cause irreparable damage to the economy in the short-run and may result in very costly adjustments in the long-run.

Thus, for example, if our future food supply proved to be grossly inadequate to meet population needs, then in sequence, its price would steadily climb, the quality of the average diet would steadily diminish, the calories per capita would decrease, the lower-income sectors of the population would suffer malnutrition and eventually starvation. While this gruesome and historically proven adjustment of demand to meet supply would be occurring, the increased price would

---

*Presented before the Subcommittee on Advanced Energy Technologies and Energy Conservation, Research, Development, and Demonstration, Washington, DC, 5 April 1977.

theoretically stimulate the diversion of economic resources to the expansion of agricultural production, so that after some years, as determined by institutional flexibility and resource availability, supply would grow until a new equilibrium of food and population would be established. Thus, the dynamics of the system always makes it self-adjusting in the long term and theoretically simple. However, the interim social implications and costs of such a catastrophe — historically ranging from illness to starvation to revolution — are so apparent that every nation plans to always have a food surplus.

If we consider energy supply in this same light — as key to the health of our economic productivity — then we face a similar dynamic sequence in the event of a major energy shortage, which might range from energy malnutrition to energy starvation. In this case, the interim social consequences are also foreseeable — higher costs of every "economic good", unemployment, lowered standard of living, recessions, etc. The result of the eventual equilibrium of supply and demand is also foreseeable — a lower average real income, shortages of goods, and a constrained lifestyle. Unfortunately, energy-supply deficiencies may become a limit to economic productivity quite rapidly, and conversely take at least 10 years to overcome by development of new supply. It is because of this time lag that the electric utility industry has urged the nation to plan for an adequate energy supply.

Within the envelope of an adequate energy supply, the substitutional relationships between various energy forms and fuel resources, and the timing of their development, is certainly an appropriate area for economic study, and classical econometric theory. However, the somewhat common notion among economists is that,because primary fuel costs to supply our energy needs are only 5% of GNP, the effect of large increases in energy costs cannot be important in the national economy. This completely misses the highly leveraged effect of energy costs on the production of all goods and services, including that of capital equipment. Omitting this feedback relationship to all other economic activity is a major error.

It is interesting that food production costs represent about 4% of the GNP. Do we consider the cost of food a minor issue? Of course not — there is no substitute — and for a large portion of our population food expenditures are relatively inflexible. The importance of energy price to the public can be seen in the recent reactions to suggestions of an additional 25¢ gasoline tax, the intervention at utility rate increase hearings, the movement for life-line rates. Perhaps the public intuition based on experience is more perceptive than elementary economic theory.

How much energy will we need in the year 2000? Two indicators give us a good idea: our anticipated standard of living and the size of our future labor force.

Historically, energy consumption and prosperity rise and fall together. A strong appetite for energy marks those periods when employment is healthy and the economy is growing. Energy consumption falls with waning employment and a sagging national economy.

The gross national product (GNP), because it measures the vast array of goods and services that are turned out by the US economy in any given year, mirrors living standards. The links that exist between energy, employment, and GNP (Fig. 1) provide a method for estimating future needs. If we know roughly the number of people working at some future date and the quantity of goods and services they need to produce, we have a basis to project the energy supplies required to support production.

The historical growth rate of the US economy over the period 1955–1975 averaged 3.5% per year in constant dollars. Accompanying this expansion was a swelling work force. The Bureau of Labor Statistics predicts a civilian work force of 119 million by the year 2000, up from 85 million today. Almost all this labor force is born by now, so future birthrates can have only a modest effect on altering its size. Based on experience, how much energy will it take to support the output of a work force this size?

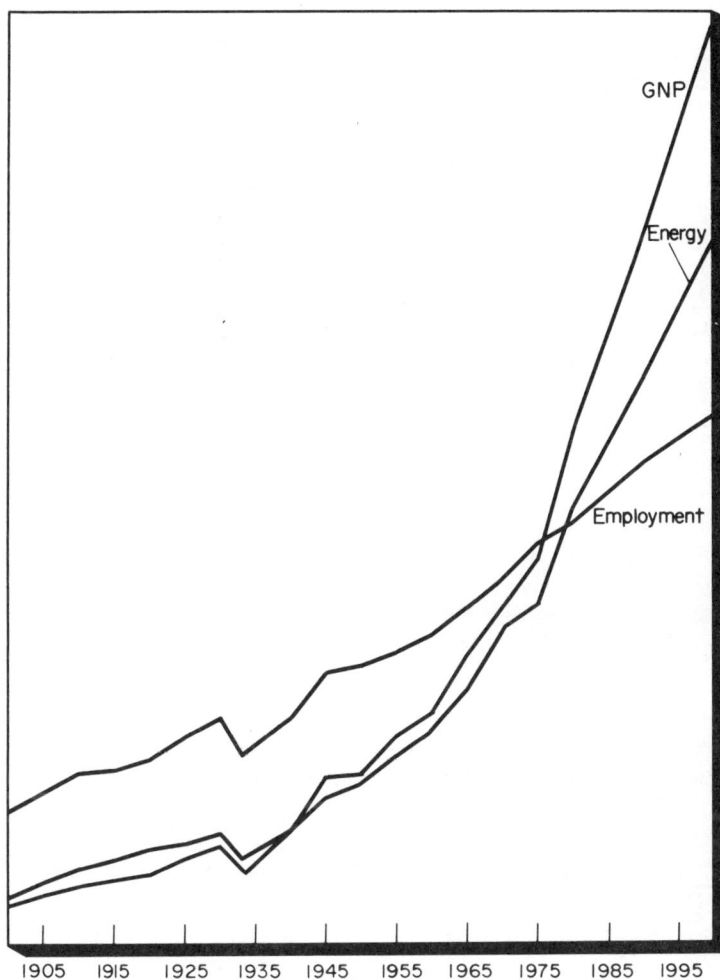

Fig. 1. Energy consumption has historically tracked employment and GNP, the two basic indicators of economic health. Between now and the year 2000, a continuation of economic growth will push energy consumption to twice present levels, even with reasonable conservation. Increased productivity per worker will balance the relatively lower total employment resulting from slowed population growth.

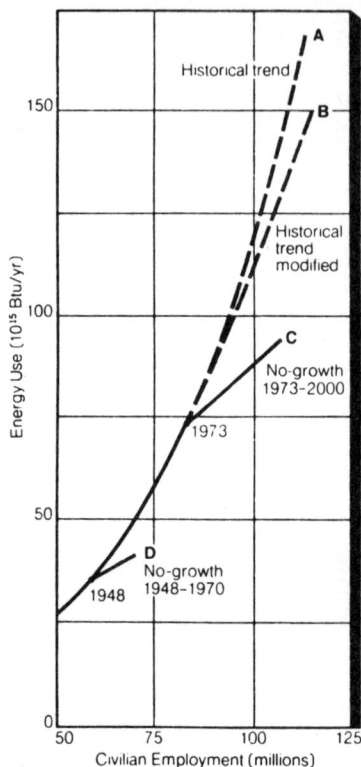

Fig. 2. Continuing the nation's historical growth trend will require far more energy than the no-growth option, which would freeze socioeconomic progress at roughly current levels (represented here by 1973 data). **A** pinpoints an energy demand of 170 quads for the year 2000, based on historical growth. **B** indicates a downward modification to allow for conservation savings of 20% and environmental costs of 10%. **C,** falling well below **A** and **B,** shows what the projected level of energy demand would be in the year 2000 if the no-growth option is chosen. **D** indicates what the no-growth option would have resulted in had it been followed from 1948 until 1970.

As a basis for technological planning, the EPRI projection of future energy needs takes into account potential changes in energy availability, technology, environmental costs, and conservation (Fig. 2). Historical economic growth would project 170 quads, conservation might save about 20%, and environmental costs may add about 10% — leading to a probable 150 quads (quadrillion Btu*)

*1 quad = 180 million barrels oil; 1 quad/yr = ½ million barrels/day.

with an uncertainty of perhaps ±10%. Thus, 150 quads — which is twice our present national energy consumption — is the amount of energy that the nation may need to provide in the year 2000 if the economy continues to grow as it has and as many think it should.

Conservation has now become a popular national concept for reducing energy demand. A distinction should be made between technological conservation which raises the efficiency with which primary fuels are converted into end-use, and sociologic conservation which requires changing national lifestyles and end-use patterns.

A detailed initial survey of the potential for technological conservation, funded by EPRI, which comprehensively covered the components of the principal end-use sectors was published in April 1976 under the title *Efficient Electricity Use*. This study is now being updated and will be reissued in several months. A condensed summation of the results is shown in Table 1, both by energy form and by using sector.

TABLE 1
Year 2000 End-use Conservation Potential Percent Savings

Conservation program	Electric energy sector	Non-electric energy sector	Total energy
None	0%	0%	0%
Reasonable	17%	25%	20%
Extreme	34%	50%	40%

Year 2000 reasonable energy-conservation potential

	Sector importance (% of total energy)		Potential energy savings (% of year 2000 sector energy demand)	Weighted potential annual savings (% of year 2000 unperturbed total energy demand)
	1973	2000 (unperturbed)		
Residential and commercial	37	40	15	6
Industrial	34	40	15	6
Transportation	29	20	40	8
	100%	100%		20%

The realism of the above judgment is supported by an independent study by Hirst and Carney, *Analysis of Federal Residential Energy Conservation Programs*, Feb. 1977. Using a detailed engineering-economic model of residential energy use developed at ORNL, they found that such use could be reduced 15% in the year 2000 by the proposed federal conservation programs and perhaps as much as 25% by a much more intensive program for applying known technologies. If this higher savings could be implemented for both residences and commercial buildings, it

FAMILY INCOME—1948 AND 1970

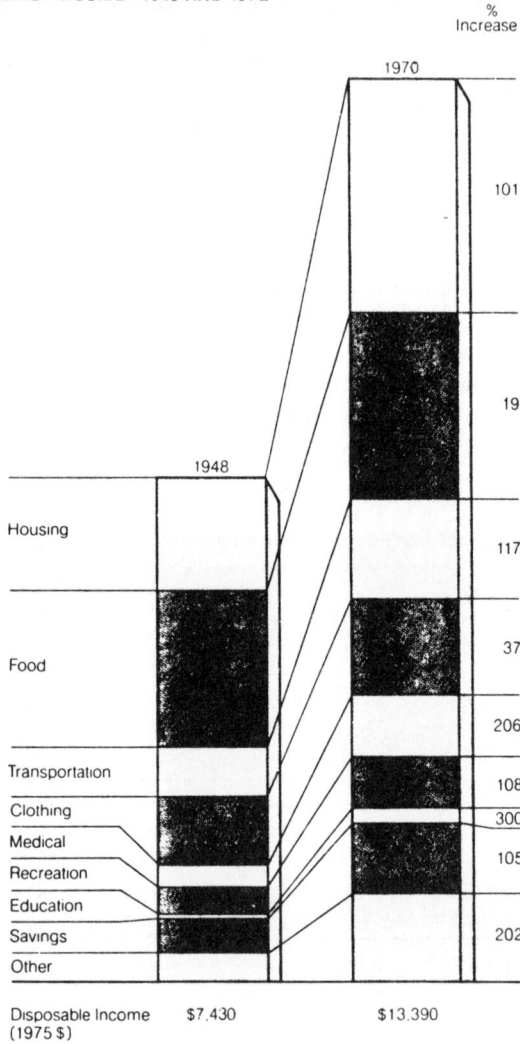

Fig. 3. Between 1948 and 1970 disposable family income nearly doubled in real purchasing power. Consumers were able to buy more—especially in areas such as recreation and education—and still increase their savings. Americans thus enjoyed substantial growth in economic well-being during this period.

would increase the EPRI projection of a reasonable 20% overall conservation to 24%. For planning purposes, we believe the 20% projection is a prudent expectation since the public acceptance of a very intensive conservation program is uncertain.

A minimal energy demand would result from a no-growth situation which would freeze the ratio of employment to energy use at its present level — that is, keep fixed the energy consumed per employed worker and fix the mix of industrial, commercial, domestic, recreational, and other energy-related activities. Freezing this employment-energy ratio would be tantamount to maintaining all major social and economic patterns, only replicating this mix in the future to accommodate an increased labor force.

What this no-growth option would really mean becomes clearer if we compare the nation's actual experience in the year 1970 with what 1970 would have been like had our technology and economic productivity been frozen shortly after World War II — say, in 1948.

The no-growth scenario would replicate the 1948 economy in an expanded form for a population that had grown from 147 to 205 million. In this hypothetical situation, the social structure and economic mix for 1970 would have been the same as those that existed in 1948. These have been described in the US Department of Commerce's monumental study, *Social Indicators 1973*. Thus, the actual changes that did occur between 1948 and 1970 can be considered a measure of the effect of increased productivity per worker, associated with increased energy input into the economy. (It is important to recognize that increased productivity per worker is associated with the continuing development and use of energy-converting systems that can produce more goods and services, so that increased productivity generally brings with it increased energy need per unit of output.)

Actual energy use in 1970 totaled 68 quads, as compared with what might have been a 1970 level of only 41 quads under such no-growth conditions. That is, the 1970 work force could have been supported with an energy input 60% as large as was actually used if the nation had been willing to live in 1970 as it did in 1948. What, then, did the nation get in return for this actual 1.66 multiplication of energy consumption per employed worker between 1948 and 1970?

A glance at Fig. 3 shows that the consumer's paycheck actually bought much more in 1970 than it did in 1948, reflecting a substantial rise in the US standard of living. A 66% increase in energy consumption per employed worker supported a 90% increase in real family income. Table 2 shows a more complete analysis for 1973. The percentage of personal income required to furnish necessities such as food and clothing declined, whereas purchases in discretionary areas such as medical care, education, and recreation increased substantially. The rise in mean family income moved a fifth of the population out of the poverty bracket to a level above the threshold of acceptable nutrition and housing.

These economic changes triggered a profound social impact as well. One example was the massive entry of women into the work force. Had the percentage of women in the work force been frozen at the 1948 level (28%, as compared to the 1970 actual of 38%), 7.8 million fewer women would have held paying jobs in 1970.

The year 1970, then, would have looked quite different had we opted for a no-growth economy in 1948. But does the same logic apply to a growth versus no-growth decision in 1976?

The future levels of GNP, productivity, and lifestyle are not bound to continue along historical lines, and other energy analysts have drawn alternative scenarious of energy supply and demand based upon different assumptions of the future performance of these economic variables. Also subject to uncertainty are the

TABLE 2

Family Income and Expenditures and Government
Expenditures (1948–1973)
(1975 dollars)

	1948 Actual or 1973 assuming no productivity growth from 1948	1973 Actual	Ratio of 1973 to 1948
Mean family income	$8250	$17,440	2.1
Taxes	820	2500	3.0
Disposable family income	$7430	$14,940	2.0
Personal consumption expenditures and savings			
Housing	$1720	$3860	2.2
Food	2290	3050	1.3
Clothing	1040	1580	1.5
Transportation	690	1670	2.4
Medical care	300	1030	3.4
Recreation	380	880	2.3
Personal business	230	760	3.3
Education	50	220	4.4
Savings	530	1220	2.3
Other	200	690	3.5
Total	$7430	$14,940	
Government expenditures per family Social security, welfare, and other transfer payments	$680	$1830	2.7
Health	80	620	7.8
Education	280	1140	4.1
Housing	20	140	7.0
Total	$1060	$3730	3.5

potential for conservation, the energy costs associated with environmental protection, and the availability and performance characteristics of energy-supply options currently under development. Perhaps the greatest uncertainties result from projections of consumer preference in future usage which can significantly influence energy demand.

Making predictions of future energy-consumption patterns in the United States has become something of a national pastime, and a prediction can be found to suit virtually every preference and sociologic outlook. Some of the recent energy forecasts will be compared, to bound the range of assumptions and energy predictions that have been made. All comparisons will be projected to the year 2000.

Generally, most studies suggest a planning projection rather than a precise forecast. The high demand projections are usually based on continuation of historical economic growth trends and are similar to the EPRI projection presented here. (See,for example; *United States Energy Through the Year 2000* (revised), by Dupree and Corsentino, Bureau of Mines, US Department of the Interior, Dec. 1975.)

The recent report, *Economic and Environmental Implications of U.S. Nuclear Moratorium 1985–2010,* Oak Ridge Associated Universities' Institute for Energy Analysis, September 1976, projects much lower energy consumption, principally due to assumed slower economic growth and relatively more conservation; the use of nuclear power and coal is predicted to grow nearly as rapidly as in the historical growth cases, but less oil and gas would be used.

The third projection is that described in the testimony of Drs. Frank von Hippel and Robert H. Williams before the Nuclear Regulatory Commission Hearing Board on the subject of Generic Environmental Statement on Mixed Oxide Fuel. This testimony was titled *Nuclear Energy Growth Projections (A Preliminary Analysis),* from the Program on Nuclear Policy Alternatives, Center for Environmental Studies, Princeton University. This forecast is quite similar to the Institute for Energy Analysis (IEA) report, but projects slightly lower energy consumption due to relatively lower GNP and higher conservation.

A fourth proposal, "Scale, Centralization and Electrification in Energy Systems", a discussion draft, and "The Road Not Taken?", in the October 1976 *Foreign Affairs* by Amory B. Lovins, is not an energy forecast, but rather a mixed sociologic and technical scenario recommended by the author which entails the curtailment of central station electrical plants, elimination of nuclear power, and rapid development of localized "soft" technologies. This is philosophically similar to the approach to social change expressed by Schumacher in *Small Is Beautiful,* Perennial Library, Harper & Row, 1975 (originally published 1973) and recommends a lifestyle pattern for the nation far less consumption oriented than that which is now being pursued.

A fifth study is the 1974 Ford Foundation publication *A Time to Choose.* This study involved a variety of approaches described as "historical growth", "technical fix", and "zero energy growth". Because of its differing approaches, it is somewhat more difficult to analyze and compare with the methodology used in this paper.

These latter four papers make policy recommendations which imply that national policies should be adopted which exert varying pressures on future energy supply and demand to fit into their predicted pattern. Such policies may be socially restrictive and may also place the nation at serious risk by creating a planned deficiency in energy supply. This issue will be discussed more fully at the conclusion.

All of these papers are more or less correct in estimating energy requirements consistent with their assumptions. The assumptions — which relate to economic growth, the range of conservation, population and employment, and the future availability, cost, and distribution of energy sources — are therefore the key features in a prediction of future energy use. It is these assumptions and how they relate to each author's view that will be examined and compared below. A more complete analytic discussion of these comparisons is contained in Appendixes A and B to this document — only the summary follows.

Table 3 compares the principal assumptions of each forecast, and the major

TABLE 3

Analysis	Projection	Employment		(1) Productivity growth rate (% Yr)	GNP		(2) % Energy Conservation	% Energy increase from Environmental improvement	Yr 2000 energy (quads)	(3) Yr 2000 electric fraction
		1975-2000 growth rate (%/Yr)	Ratio (2000/1975)		1975-2000 growth rate (%/Yr)	Yr 2000 (10⁹ 1958$)				
EPRI	Historical productivity growth	1.36	1.40	2.28	3.64	2025	20	10	149	.55
	No product growth	1.36	1.40	.13 (4)	1.49	1183	20	10	89	*
IEA	High case	1.24	1.36	2.0	3.24	1834**	15.7	0	126	.51
	Low case	1.18	1.34	1.88	3.06	1753**	31.0**	0	101	.47
von Hippel-Williams	Status quo	1.24	1.36	1.7	2.94	1701**	21.2*	0	113	*
	Moderate	1.24	1.36	1.7	2.94	1701**	37.5**	0	89	.40
Lovins	*	*	*	*	*	*	*	*	95	.16
Ford	Historical	1.43**	1.43**	1.94**	3.37***	2366	0	0	187	.40
	Technical fix	1.43**	1.43**	1.78**	3.21***	2277	35	0	123	.25
	ZEG	1.43**	1.43**	1.79**	3.22***	2283	47	0	100	.31

Notes: (1) Productivity growth rate for GNP/employee-year. (2) Computed from base case. (3) Fraction of primary fuels used for electricity production. (4) Based upon a return to 1973 productivity. (*) Not given. (**) Computed by EPRI to be consistent with other economic assumptions of authors. (***) Computed from Ford Foundation estimate of 1975 GNP of 1020 billion 1958$, actual 1975 GNP 815.7 billion 1958$.

EPRI estimates were based upon a continuation of growth from 1973 to 2000. From 1973 to 2000 the growth rates were to be: employment 1.27%/yr, productivity 1.99%/yr, and GNP 3.26%/yr.

differences are apparent. Small differences in the assumed annual rate of increase in the key factors, when compared to the year 2000, make very large differences in GNP and energy-demand outcomes. Except for the potential of technological conservation (improved energy efficiency), all the other factors do affect the total economic output of the nation, and thus the cost and availability of goods and services.

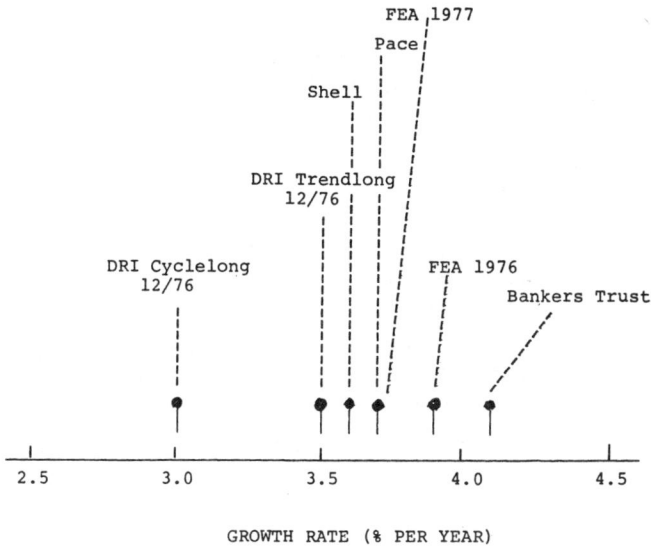

Fig. 4a. Long-term forecasts to 1990.

It is pertinent to compare these forecasts of economic output based on energy-demand-related factors to the economic growth targets set by the national goal of full employment of the labor force. In this domain of conjecture, the spread of estimates is as wide as are those from the energy sector. Figures 4a and 4b list some of the projections assembled from a variety of sources. Again, it should be realized that small differences in annual growth rates have a large compounded effect by the year 2000. For example, a 1% shift in average annual growth rate from 2.75% to 3.75% produces an increase in the year 2000 GNP of 28%. Obviously, there should be some consistency in our planned economic growth and planned energy growth.

In addition to the sociologic and institutional uncertainties that will influence our economic and energy futures, there are other major uncertainties that could substantially change the course of events. Among these are fuel availability, both domestic and foreign, environmental impacts and the technology of their control, and end-use technical changes that shift the demand mix (for example, the electric automobile). Such factors may have a much greater influence on long-range outcomes than those planning factors presumably under our control.

Under such circumstances of great uncertainties, it is apparent that it is not feasible to predict a cumulative energy demand accurately enough to permit the choice of a well-defined path. Unfortunately, the potential risks of various energy policies are quite different — it takes about 10 years to rectify a shortfall in energy supply. On the other hand, it is hard to conceive of a domestic energy surplus given our currently increasing dependence on foreign oil. At worst, if such a surplus were to occur, it would mean leaving some facilities under-utilized for a period of time until demand equates with supply. And, in the case of electric power generation, a "surplus" of coal- and nuclear-fired power plants would permit the more rapid retirement of existing oil- and gas-fired facilities.

Thus, the calculable capital cost of a hypothetical over-expansion of supply should be weighed against the very real social and economic costs of an energy deficiency — such as reduced economic growth, unemployment or reduced average income, the increased cost of goods, and impaired international trade — or alternatively, increased bondage to foreign oil, if it is then available.

If a prudent national policy is to avoid the coercive pressures of restricted supply, then what should we do to minimize such a risk? The oft-heard answer is the prudent litany: technological conservation to improve end-use efficiency; develop all supply options within the framework of a planning target that provides for future flexibility; continuous and intensive assessment of resource, economic, environmental, and social trends, both domestic and international.

To many of us actively engaged in the national energy scene, this is a fast changing period and a time when we should keep all options open and under development.

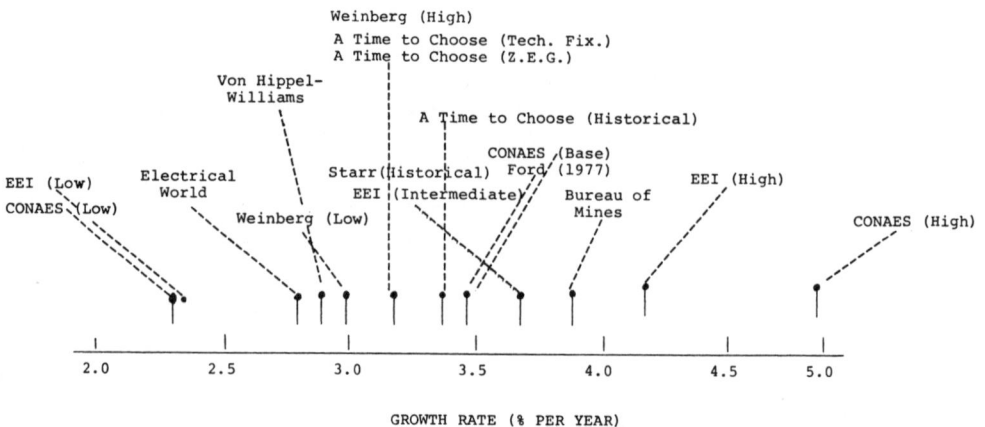

Fig. 4b. Long-term forecasts to 2000.

Appendix A

Summary of Energy Forecasts

(EPRI, Institute for Energy Analysis, von Hippel and
Williams, and comments on the Ford Foundation report)

All of these forecasts can be represented as using the same methodology:

(1) Employment, productivity estimates produce GNP estimate.
(2) Historical GNP/energy provides a base case.
(3) Modifications for conservation (and environmental energy in the EPRI
report) give the final estimate.
(4) Methodology for the electric fraction varies, and the assumptions which
lead to the estimate of electric fraction can not be compared — frequently
the fuel mix/electric fraction forecast is given without derivation.

Employment Assumptions

EPRI — projects employment and average work week.

Ford Foundation — projects man-hours (equivalent to EPRI).

Because of the 1974–1975 recession, EPRI estimates are computed from 1973
(see Tables 4 and 5).

IEA, von Hippel — projects full-time equivalent (this measure is calculated
on a year-by-year, industry-by-industry basis and does not include declining
work week effects). See Table 6.

Three measures are used which make direct comparison difficult (see Table 7).

TABLE 4

EPRI	1973	2000	Ratio 2000/1973	Annual growth rate
Employment	84.4×10^6	119×10^6	1.41	1.27%
Work week	39 hr	36 hr	.92	−.30%
Man-hours*	3.292×10^9	4.284×10^9	1.30	.975%

*These are not the total man-hours worked because of part-time employees included in
the total employment data. It is assumed that the full-time to part-time mix will not
change appreciably and that the growth rate for man hours worked will be valid.

TABLE 5

Ford Foundation	1975	2000	Ratio	Annual growth rate
Employment	84.4×10^6 (1973)	121.2×10^6	1.436	1.34%
Man-hours				
Historical growth	173.115×10^9	262.557×10^9	1.52	1.67%
Technical fix	173.115×10^9	266.548×10^9	1.54	1.73%
Zero energy growth	173.115×10^9	271.274×10^9	1.57	1.80%

Note: ZEG scenario seems to imply that average work week will increase from 39 hours to 43
hours per week.

166

CURRENT ISSUES IN ENERGY

TABLE 6

	1975-85	1985-2000	2000/1975 ratio	Average growth
IEA High case	1.9%	.8%	1.3634	1.24%
IEA Low case	1.9%	.7%	1.3431	1.18%

TABLE 7
Summary

	Average growth rate (1975-2000)
Ford ZEG man-hours	1.8%
technical fix man-hours	1.73%
historical man-hours	1.67%
employment	1.43%
EPRI man-hours (1973-2000)	.975%
employment (1973-2000)	1.27%
IEA High case, employment	1.24%
Low case, employment	1.18%
von Hippel-Williams employment	1.24%

Productivity

The Ford Foundation makes no productivity estimates, but they can be computed from GNP estimates. IEA and von Hippel-Williams calculate productivity by GNP/full-time equivalent employee-year. EPRI calculates productivity by GNP/employee-hour, in which employee-hours are calculated by average work week times number employed. (This over-estimates man-hours but should give accurate growth rates.) (See Table 8.)

Note on productivity. The historical GNP/employee-hour data has demonstrated year to year fluctuations (it was comparatively high during World War II and low during the recession of 1914–1917, the Depression, and just after World War II). In all cases to date, the productivity has returned to the extrapolated trend. Basing the historical trend on growth between 1945 and 1975 would be very misleading because 1945 productivity was extremely high relative to the neighboring years, and 1975 was a recession year. To demonstrate the sensitivity, the GNP/employee-hour is given for several periods.

Annual growth
1945–1975	1.74%
1946–1973	2.57%
1965–1975	1.38%
1963–1973	2.36%

TABLE 8

EPRI		1973-2000	1975-2000
GNP/employee-hour		2.29%	2.60%
GNP/employee-year		1.99%	2.28%

IEA GNP/employee-year	1975-85	1985-2000	1975-2000
High case	1.7%	2.2%	2.0%
Low case	1.7%	2.0%	1.88%

von Hippel-Williams	1975-85	1985-2000	1975-2000
GNP/employee-year	1.4%	1.9%	1.7%

Ford GNP/employee-hour	1975-85	1985-2000	1975-2000
Historical	1.89%	1.57%	1.7%
Technical fix	1.63%	1.31%	1.48%
ZEG	1.60%	1.31%	1.42%

GNP Forecasts

Combining the employment-productivity estimates of growth results in a GNP growth forecast. The results are shown in Table 9.

TABLE 9

	Growth rates of GNP		
	1973-2000		1975-2000
EPRI	3.26%		3.64%

	1975-1985	1985-2000	1975-2000
IEA high	3.6%	3.0%	3.24%
IEA low	3.6%	2.7%	3.06%
von Hippel-Williams	3.3%	2.7%	2.94%
Ford*			
Historical	3.58%	3.22%	3.365%
Technical fix	3.42%	3.07%	3.21%
ZEG	3.42%	3.09%	3.22%

*Ford estimates from 1975 GNP estimated at 1020 billion 1958$, actual 1975 GNP 815.7 billion.

GNP estimates, 10^9 1958 dollars	2000
EPRI	2025
IEA high case	1834
IEA low case	1753
von Hippel-Williams	1701
Ford	
Historical	2366
Technical fix	2277
ZEG	2283

GNP-Energy

The historical GNP-energy relation is:

$$E_n = .0807 \text{ GNP} + 5.495$$

Corresponding to this equation the computed and stated base case energies are as shown in Table 10.

TABLE 10

	Computed	Stated	Final estimate
EPRI	168.9	~170	~150
IEA			
High	153.5	149.3	125.9
Low	147.0		101.4
von Hippel-Williams	142.8		89.2–112.5
Ford			
Historical	196.4	187–188	
Technical fix	189.2		123–124
ZEG	189.7		100

Conservation

Conservation is computed from base case energy when given; if none is given it is computed from the historical energy corresponding to the estimated GNP (Table 11).

TABLE 11

	Reduction from base case
EPRI	20%
IEA	
High	15.7%
Low	31%
von Hippel-Williams	
"Moderate"	37.5%
"Status quo"	21.2%
Ford	
Historical	0
Technical fix	35%
ZEG	47.3%

The von Hippel-Williams "status quo" case is defined as one in which energy prices remain stable at 1975 levels and no new conservation measures are enacted beyond auto fuel economy standards already established.

Only the EPRI estimates include an environmental energy-cost estimate (of 10%) to compensate for efficiency losses which occur due to more stringent environmental standards. This figure also compensates for the expected leveling of heat rate (which in the past was declining).

Electric Fraction

TABLE 12

EPRI	54.8%
IEA	
High	50.8%
Low	46.6%
von Hippel-Williams	
"Moderate"	39.8%
"Status quo"	not given
Ford	
Historical	39.5–42.5%
Technical fix	25%
ZEG	31%

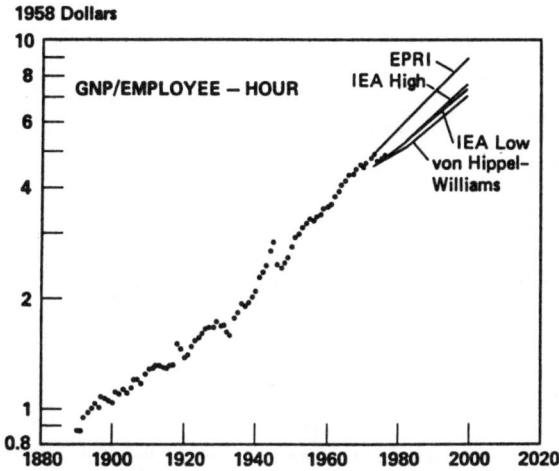

1958 Dollars

Fig. 5. Productivity.

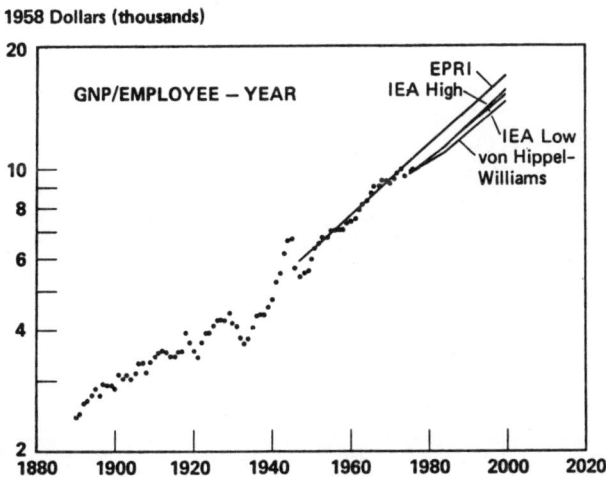

1958 Dollars (thousands)

Fig. 6. Productivity.

Appendix B

Methodology and Analysis

The strong historical relationship between GNP, employment and energy is usually used as the basis for projections of future energy requirements. Figures 7, 8, and 9 show the high degree of correlation which exists between these three factors. Several factors are expected to modify this historical relationship, factors that are of such recent origin that their influence is not yet reflected in historical data. The first of these is the impact of rising energy prices. The obvious result expected is a conservation of energy relative to that which otherwise would have been consumed. The second is the additional energy required to achieve environmental quality goals.

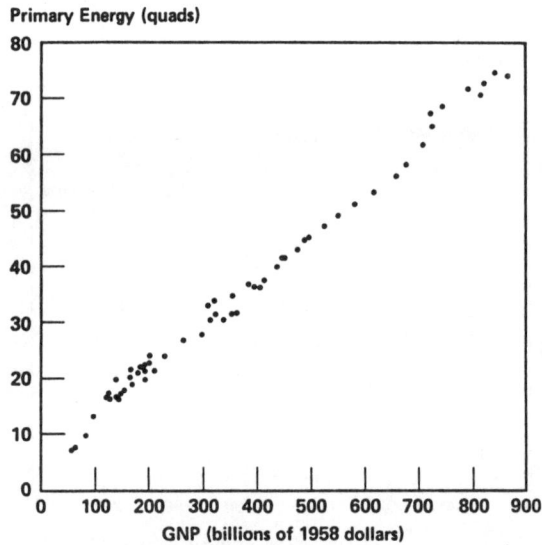

Fig. 7. Energy versus GNP.

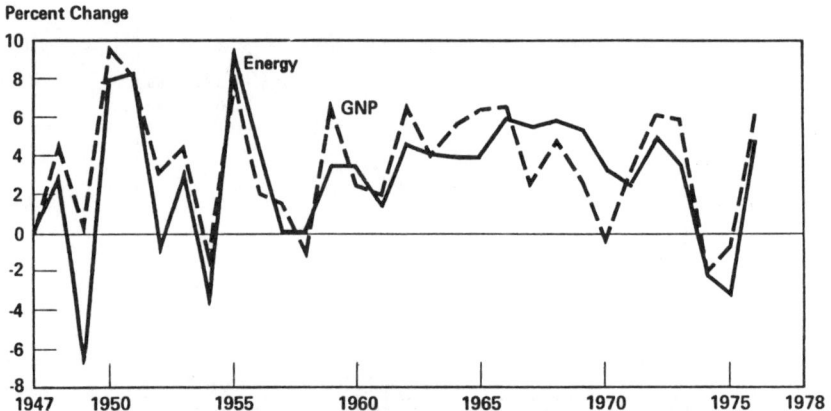

Fig. 8. Changes in energy and GNP.

Energy (quads)

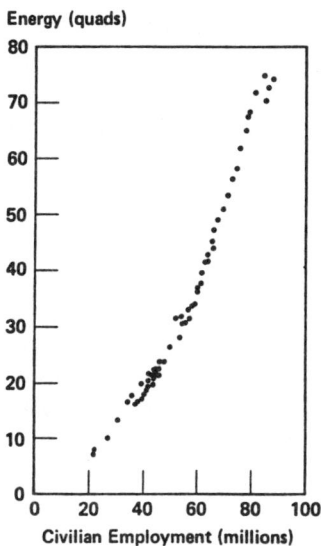

Fig. 9. Energy-employment.

The IEA (Institute for Energy Analysis of Oak Ridge) forecast uses the GNP-energy relationship but includes several other factors such as government intervention to promote conservation, the effect of changing age distribution within the population, changes in lifestyle and taste, and saturation of energy-using devices.

The paper by von Hippel and Williams uses a GNP-based methodology but takes a different approach for conservation. Energy end uses are identified and specific conservation measures for each end use are discussed. The energy forecast is the sum of these requirements.

Lovins does not use a GNP based methodology in his scenario. His comment on this topic is:

> "Since the energy needed today to produce a unit of GNP varies more than 100-fold depending on what good or service is being produced, and since GNP in turn hardly measures social welfare, why must energy and welfare march forever in lockstep?"
> Amory B. Lovins,
> "The Road Not Taken?", Section X,
> *Foreign Affairs*, Oct. 1976

Year 2000 GNP

To project year 2000 GNP the following identity will be used:

$$GNP = L \cdot H \cdot P$$

in which L is employment, H is hours worked, and P is productivity defined as GNP produced per employee hour. The IEA paper and von Hippel-Williams use a similar method, except that productivity is based upon output per equivalent full-time employee per year, and hours worked are not considered.

Projections of year 2000 labor force are not particularly controversial because the vast majority of people who will be working at that time have already been born, and the estimate is quite insensitive to different projections of population growth. Those trends which should be considered are the changing age distribution of the population, the trend towards earlier retirement, and the increasing percentage of women joining the labor force. The Census Bureau projections for year 2000 population aged 16 and over are 199.7 million (Series II) and 194.7 million (Series III). Taking 197 million as a reasonable estimate, and using a participation estimate of .63 (the same as that used in the IEA report), the labor force would be 124 million. The IEA report estimates 123 million for their low case, 124 million for their high case. As a basis for planning, 4% unemployment is assumed (it is considered unlikely that unemployment will be less than this value). The year 2000 employment corresponding to these assumptions is 119 million.

The historical data for the hours in the average work week is shown in Fig. 10. Aside from the peaks associated with the Depression and World War II, the average hours worked per week have been slowly declining. An extrapolation of this trend projects a 36-hour work week for the year 2000, down from the 39 hours at present. This work-week trend is not included in the IEA or the von Hippel-Williams analysis.

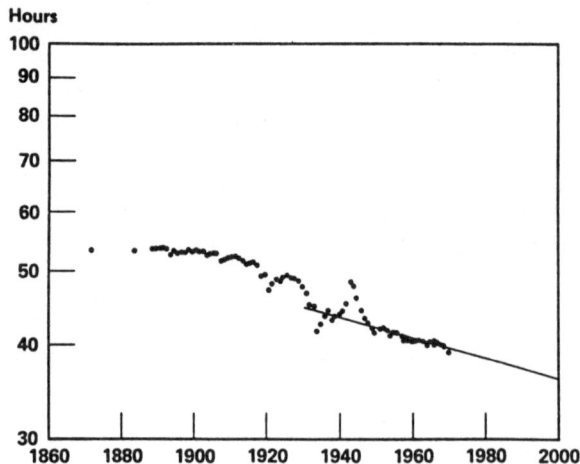

Fig. 10. Work week trend.

Productivity

The most sensitive estimate is that of future productivity; the historical trend (Fig. 11) indicates GNP in 1958$ per employee-hour. Productivity for 1974 and 1975 was lower than for 1973, and as a result projections vary depending upon whether an extrapolation from 1973 or 1975 is taken. Short-term productivity losses seem to be recovered rapidly (see, for example, the data for 1922, 1934, and 1950). As a result, it is assumed that the lower rates of the past 2 years will not affect the long-term growth. The relatively smooth growth in the 20-year period 1953–1973 gives an annual growth rate of 2.3% and this is the value used in the EPRI projection. A continuation of this trend to 2000 would result in a production of 9.09 1958$ of GNP per employee hour versus 4.90 1958$ in 1973. To reach this productivity from the 1975 level will require a growth rate of 2.6% per year. While recovery in productivity following the recession of 1974–1975 is by no means assured, it is comforting to note that three times in the past recessions have been followed by a rapid return to historical trends.

Fig. 11. Productivity.

In addition, preliminary 1976 data indicates that real GNP grew 6.1% from 1975 with employment increases accounting for about 3.2% of the growth and productivity accounting for 2.9%. The 1976 productivity is back roughly to the 1973 level; the increase from 1975 to 1976 is at a higher rate than the historical trend.

Using the formula for GNP defined previously, the year 2000 EPRI planning projection (labor × hours × productivity) is $2025 billion in 1958$.

This productivity growth rate (and subsequent GNP estimates) is substantially higher than that used in the formulation of the IEA and von Hippel-Williams forecasts. Because a different basis is taken the productivity growths may not be directly comparable; these studies compute productivity as GNP per full-time equivalent worker-year. The historical data used by IEA indicates a rate of productivity growth of .9% from 1965 to 1975, and 2.2% from 1945 to 1965. For the same periods, productivity as defined in this analysis grew at 1.3% and 1.9%, respectively. An examination of the productivity data with extrapolations (Figs. 11 and 12) indicates that 1945 and 1975 are highly unrepresentative years on which to base a projection. The 1945 productivity was at a record high relative to the times, due principally to the wartime effort. This level was not exceeded until the early 1950s. The 1975 productivity was relatively low as it has been in past recessions (1946–48, 1932–33, and 1917). The IEA analysis assumes that productivity (measured in GNP per full-time equivalent worker per year) will grow at 1.7% per year until 1985, then grow at 2.0% to 2.2% per year until the year 2000. This growth occurs from the 1975 point, without a rapid recovery of the 1974 and 1975 losses.

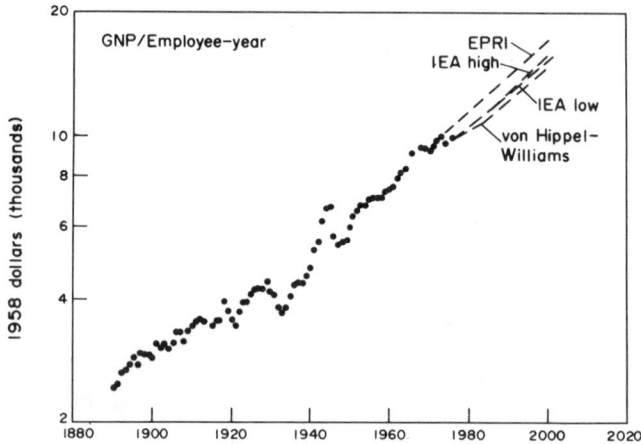

Fig. 12. Productivity.

The von Hippel-Williams productivity projection, based largely upon the IEA report, is even lower. They predict GNP per full-time equivalent employee to grow at 1.4% until 1985, then at 1.9% until the year 2000, giving an average growth of 1.7% for the 25-year period. Although both these reports project productivity growth to slow from the previous two decades, they both term their estimates conservatively high.

These economic assumptions are summarized in Table 13.

TABLE 13
Employment Assumptions

	1975	2000	Ratio 2000/1975	Annual growth rate
EPRI Employment	84.8×10^6	119×10^6	1.41	1.35%
	1975-80	1985-2000		
IEA high annual rate	1.9%	.8%	1.3634	1.24%
IEA low annual rate	1.9%	.7%	1.3431	1.18%
1975-85	1985-90	1990-2000		
von Hippel annual rate 1.9%	1.0%	.7%	1.3634	1.24%

TABLE 14
Annual Productivity Assumptions

EPRI		1973-2000	1975-2000
GNP/employee-hour		2.29%	2.60%
GNP/employee-year		2.0%	2.28%
IEA GNP/employee-year*	1975-85	1985-2000	
High case	1.7%	2.2%	2.0%
Low case	1.7%	2.0%	1.88%
von Hippel-Williams			
GNP employee-year*	1.4%	1.9%	1.7%

*On a full-time equivalent basis.

TABLE 15
GNP Estimates

	Growth rates		
	1973-2000		1975-2000
EPRI	3.26%		3.64%
	1975-1985	1985-2000	
IEA high case	3.6%	3.0%	3.24%
IEA low case	3.6%	2.7%	3.06%
von Hippel-Williams	3.3%	2.7%	2.94%
	10^9 *1958 dollars* 2000		
EPRI	2025		
IEA high case*	1834		
IEA low case*	1753		
von Hippel-Williams*	1701		

*Calculated from stated GNP growth rates and the 1975 GNP of 815.7 × 10^9 1958$.

Energy Planning Target

A linear extrapolation of the historical energy-GNP data (Fig. 13) predicts that 168.9 quads (1 quad = 10^{15} Btu) of primary energy would be consumed for a year 2000 GNP of 2025 billion 1958$.

Base case energy consumption may be calculated from the excellent linear fit between GNP and energy. The equation is:

$$E_n = .0807 \text{ GNP} + 5.5$$

in which energy is measured in quads and GNP in billions of 1958$. The IEA GNP estimates calculated from the stated growth rates would yield base case energy estimates of 153.5q for the high case and 147q for the low case. The stated base level for the higher case is 149.3; no base for the low case is given. The base case calculated from the von Hippel-Williams GNP forecast is 142.8 quads.

As was mentioned previously, two significant factors are expected to alter the historical relationship between energy and GNP; these are conservation (as affected by energy cost) and environmental energy costs.

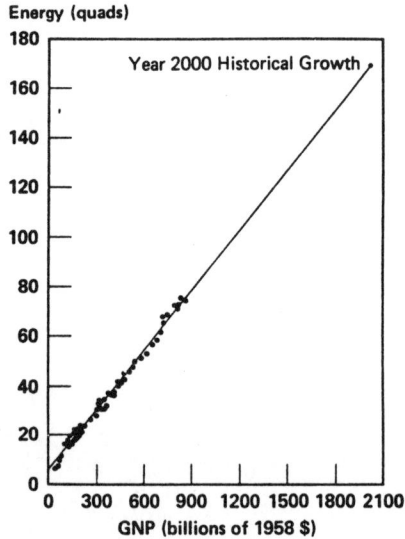

Fig. 13. Energy and extrapolation of historical trend.

Conservation

A reduction of the energy that would be required under an extrapolation of past trends can be expected due to increasing energy prices. In addition, it is anticipated that conservation will occur due to legislation requiring that specified efficiencies be met in energy-consuming devices. Mandatory gasoline mileage targets have already been established, many states have established or increased insulation requirements for new housing, and a recent law will require an increase in appliance efficiency of 20% by 1980. All of these actions probably fall within the category of cost-effective conservation that probably would occur eventually. In essence, these regulations tend to increase the rate at which adjustment to rising energy costs occurs.

In addition to price-induced conservation measures, voluntary conservation (as from following a conservation "ethic") can occur. While the various actions within this category — reduced thermostats and the formation of car pools for example — do save money as well as energy, the primary incentive is based on an appeal to patriotic motives. While such measures seem effective under emergency conditions — the 1973 oil embargo and the gas shortage during the winter of 1976–77 — the long-term potential of this type of conservation is not known. Conservation of this type is not included in the EPRI planning target.

In response to higher energy prices, short-term action will be taken to cut energy costs through the use of capital expenditures. Most short-term action will be limited by the existence of residences, commercial buildings, industrial plants and automobiles that have been built based on the economics of inexpensive energy. Over the long run, as new capital stock is designed and manufactured, the impact of energy conservation will become more significant. Because it takes decades to replace our capital stock, energy conservation will be slow to take effect.

It should be noted that there is a large difference between the conservation level that is technically possible and the level that is economically attractive. In the absence of patriotic motives or governmental regulations, the intelligent consumer will conserve energy to the point at which additional conservation costs more than additional energy.

There are a number of institutional hindrances that make particular conservation investments unattractive to the individual consumer, though attractive to society as a whole. Consider an apartment building in which the tenants pay their own fuel bills. The owner of the apartment has no incentive to invest capital that would result in energy savings, because the savings would accrue to the tenants. In addition, there are a number of cases in which cost-effective conservation could be implemented, but the cost of the bother exceeds the benefit, as is the case for the homeowner who desires attic insulation, must find a contractor, then apply for a loan to cover the cost of the work. If all this effort might produce a monthly savings of a few dollars, with two bills to pay instead of one, chances are the homeowner will not make the improvements. These extra "transaction costs" must be considered.

The availability of capital also affects the rate of adoption of conservation technology. Any particular company is limited in its ability to raise capital; sometimes a choice must be made between expansion and conservation. As a result, in some instances, cost-effective conservation investments will not be made.

Uncertainty is another factor, since in times of uncertainty, people prefer to have their money in the bank rather than tied up in capital equipment. Uncertainty about future prices of fuel and other materials affects both the homeowner considering additional insulation, and the industrial manager considering refitting his plant or replacing machinery. In short, "frictional" forces in the economy — institutional patterns, the capital stock, and transaction costs — will reduce the optimal amount of energy conservation.

The optimal amount of conservation from a price standpoint is not known, and estimates vary widely. It is probably fair to say that most estimates of this lowest cost point fall in the range of 20 to 50% from the historical extrapolation. Given the imperfections of the market, the uncertainties, and the institutional barriers to capital investment for energy conservation, 20% conservation is a reasonable amount for a planning target.

From the base historical case of roughly 170 quads, the 20% conservation estimate would reduce the year 2000 energy planning target to 136 quads. For a comparison, the IEA report uses a conservation estimate of about 31% by the year 2000 for the low case, and 15.7% for the high case, both relative to the base case energy that would be consumed if the GNP-energy relationship were simply extrapolated.

   The analysis of von Hippel and Williams also includes two different conservation cases. Their most optimistic estimate (termed moderate conservation) is for
37.5% conservation from the base case computed previously. The second case
(*status quo*) projects a 21.2% reduction. This second case is defined as one in which
energy prices remain stable at 1975 levels and no new conservation measures
beyond the auto fuel economy standards are enacted. For the moderate conservation case energy prices are assumed to rise 2% annually in constant dollars, and
new conservation initiatives are assumed to develop.

   Lovins' projections of conservation refer to improvements in end-use efficiency
rather than reductions from an historical base case. The two papers by Lovins
mentioned earlier cite quite different projections. In the *Foreign Affairs* paper he
cites an American Institute of Physics report:

> "Theoretical analysis suggests that in the long term, technical fixes *alone* in
> the United States could probably improve energy efficiency by a factor of at
> least three or four."

He cites a second reference which suggests that efficiency in the year 2000 could
be almost double that of today. In his Oak Ridge discussion draft these prospects
are lowered to a projection of doubling efficiency over the next 50 years. An
excellent analogy describing the pitfalls of relying on theoretical analysis of the
type mentioned above was offered by Bert Wolfe:

> "To take this report and argue that it suggests that we could improve energy
> efficiency by a factor of 3 or 4 is like suggesting that we will, in the future,
> travel between New York and Los Angeles in a second because the laws of
> physics say that we can approach the velocity of light as a theoretical limit."

Precisely the difference between what is theoretically possible and what is
practically possible with today's technology led to the EPRI estimate used in this
analysis.

   The method used to estimate conservation by von Hippel and Williams has one
substantial drawback: by calculating savings on a case-by-case basis, they neglect
the possibility that new uses for energy will develop in the balance of the century.
With real per capita income expected to nearly double an estimate of future
consumption patterns for all goods, energy included, must be suspect. This
growth in income provides a tremendous range in lifestyle choice — and von
Hippel and Williams have examined only one of a number of possible lifestyles.
By taking credit for every conceivable future development which might decrease
energy use, and ignoring those possible factors which could increase consumption, the von Hippel-Williams estimate serves as an upper bound on possible
conservation.

   The spread in the conservation estimates between these reports is not surprising, given the differences in the basic goals of the analyses (IEA and von Hippel-Williams have attempted to estimate a likely value for future demand, the
EPRI analysis seeks to provide a conservative basis for planning). Given this
intent, 20% conservation seems quite reasonable as an assumption for planning.

Factors Which Could Increase
Energy Requirements

Several factors which could increase the demand for energy are of such recent origin that they are not adequately reflected in the historical GNP-energy relationship. Foremost among these are the energy requirements associated with environmental protection.

The achievement of societal objectives for an improved environment is often accompanied by added inefficiencies in the use of fuel and capital resources. The social benefits of environmental improvement should justify such additional costs, and estimates of future demand on resources should take them into account.

Consider, for example, the decrease in conversion efficiency of conventional electric power stations which results from changing from water-cooled to air-cooled condensers — as part of a program to reduce the detrimental effects on water resources. Due to the higher temperature of the steam cycle condenser with air cooling, the fuel consumption is increased about 6–10% for fossil-fuel plants and 7–13% for nuclear plants.

The removal of undesirable chemicals from coal prior to combustion also results in increased energy use. At present, all precombustion systems for converting coal to purified gases or liquids involve chemical processes which require heat — from 20–50% of the heat content of the input coal, depending on the cycle detail. Flue gas purification required after the direct combustion of high sulfur fuels takes about 5–10% of the input heat. Purification during combustion by a fluidized bed boiler may use only about 2% of the input energy, but this is the least developed approach and probably will not make much commercial impact for several decades. All these methods add substantially to the energy and capital costs of the electric power systems.

In the non-electric sector, restrictions on emission of sulfur will increase the energy requirements when oil is burned. The expected reduction in the share of energy supplied by oil and gas could lead to a decline in the efficiency of energy use, because the substitutes for these fuels, synthetic fuels from coal and shale oil, require substantial energy inputs to achieve a fuel in a desirable form. Although coal gasification and liquefaction should reduce the environmental impact of burning coal, the proposed synthetic-fuel technologies lose from 20 to 50% of the input energy of the coal.

Implicit in the historical GNP-energy ratio is a declining heat rate for electric power plants. The heat rate of new fossil power plants is no longer decreasing, and increased use of light-water reactors (with heat rates higher than those of modern coal-fired plants) will break the long trend of declining heat rate. This is in addition to the efficiency impacts of environmentally motivated features such as scrubbers and dry cooling towers. The widespread use of storage devices by electric utilities will also increase primary fuel requirements. These factors will be discussed in more detail when the electricity planning target is estimated.

Nothing has been mentioned about the patterns of energy consumption which will correspond to the lifestyles chosen for the year 2000. This is the greatest source of uncertainty. If the economic planning target is met, per capita GNP will roughly double by the end of the century. To anticipate the energy use at that time requires an estimate of how this income will be spent — any estimate of this type must be extremely judgmental. For the purposes of this analysis it is assumed that the spending patterns continue along the historical path, with the conservation modifications lowering energy use through improvements in end-use efficiency alone.

The only quantitative change from the base case modified for conservation that will be included in the planning target is an allowance for the environmental energy uses. Based on present environmental objectives, it is reasonable to expect that total energy input requirements will be increased 10% over the otherwise anticipated demand. The modification for conservation reduced the planning target from 170 quads to 136 quads; a 10% increase for environmental effects raises this figure to our year 2000 energy planning target of about 150 quads. The potential effects of conservation and environmental energy use upon GNP are considered negligible. Environmental energy costs were not specifically included in the IEA or von Hippel-Williams forecasts although this factor may be included in their estimates of potential conservation. These energy forecasts are compared with the EPRI planning target in Table 16.

TABLE 16
Comparison of Estimates

	EPRI analysis	IEA high	IEA low	von Hippel status quo	von Hippel moderate	Lovins' scenario	No productivity growth (EPRI methodology)
2000 GNP (billion 1958$)	2.025	1834	1753	1701	1701	*	1183
Base case energy (quads)	168.9	149.3	*	*	*	*	101.2
Conservation from base case (quads)	—33.9	—23.4	*	*	*	*	—20.2
Environmental energy requirement (quads)	13.5	0	*	*	*	*	8.1
Year 2000 energy (quads)	148.6	125.9	101.4	112.5	89.2	95.0	88.9

Note: To prevent the implication of precision suggested by four significant digits, the EPRI numbers are frequently cited as 170q for the base case and 150q for the year 2000 planning target.
*Not given.

To indicate the sensitivity of these estimates to the economic and conservation assumptions, a breakdown of the differences between the EPRI case and the IEA high case is given in Table 17.

TABLE 17

Impact of Economic, Conservation, and Environmental
Assumptions on Year 2000 Energy
IEA high case vs. EPRI analysis

	Difference in quads (− indicates EPRI < IEA)	
Economic-energy		
Impact of differing productivity estimates	15.4	
Difference in computation of base case if same GNP	4.2	
Subtotal	19.6	19.6
Conservation	−10.4	
Environmental ·	13.5	
Net	3.1	3.1
		22.7q

Table 17 summarizes the reasons for the difference between the IEA high case and the EPRI planning target. Note that 68% of this difference is directly attributable to the different productivity rates assumed for the future.

Year 2000 Electricity Planning Target

The estimate of 20% savings due to energy conservation includes both the electric and non-electric sectors. It is anticipated that savings will be more easily achieved in transportation and space heating than in the use of electricity. The point of use efficiency of electricity is already quite high; as a result, the reduction in consumption from the base case will be less than the 20% average estimated for both sectors. Craig Smith, *Efficient Electricity Use*, estimates the range of potential savings to be from 17 to 34% by the year 2000. Because of the slow adjustment of capital stock, the effects of uncertainty, transaction costs, and because of the conservative nature of the planning target, electric sector conservation is assumed to reach 17% of the base case electricity requirements.

Due to this smaller potential for conservation in the electric than in the non-electric sector, fuel requirements for electrical generation are estimated from the base case of 170 quads. Subsequent modifications can then be made for electricity conservation and for energy costs associated with environmental improvement.

From 1960 to 1973 the energy input to generate electricity as a percent of total energy grew at an annual rate of 2.6% as shown in Fig. 14. An extension of this rate of increase to the year 2000 indicates that 53% of all primary energy will be used as fuel for the generation of electricity. This extrapolation appears reasonable although it is not evident how much of the past growth was motivated by cost and how much by the value of convenience, operational flexibility, uniqueness of application, and other such characteristics of electricity. While it is recognized that the growth rate of the electric fraction must eventually decrease, electricity use is still far from saturation. The slowing is not assumed to occur before the year 2000.

The rate of substitution of electricity for other energy forms could be altered by a number of factors. Certainly the most important is the anticipated future supply of oil and gas. Even today, the threat of short supplies of natural gas in the future has resulted in the choice of electric heat for much new construction. If, in the future, significant domestic supplies of oil and gas are found or early success comes to the programs to develop synthetic fuels from coal, the growth of the electric fraction could be slowed. The rate of technological developments such as the electric car will also affect this fraction. The uncertain impacts of all these factors suggest the choice of a band of outcomes for the fraction of fuels used to generate electricity of 53 ± 10% for the historical case.

Built into the historical energy and electricity consumption data is the increasing efficiency with which primary fuels are converted to electricity. This efficiency (expressed as the heat rate) is presently about 10,500 Btu/kWh. Extrapolation of recent heat rate trends gives 10,000 Btu/kWh for the year 2000, and would reflect continuing improvements in conversion efficiency.

The base case for electricity consumption corresponding to the growth of electric fraction (which measures the fraction of primary fuels used for electric generation) and to the historical heat rate trend is 9 trillion kWh (53% of 168.9 × 10^{15} Btu divided by 10,000 Btu/kWh). The planning target for electricity consumption: the base case reduced by the 17% conservation is 7500 billion kWh.

The modification for environmental energy does not affect this electricity planning target; the additional energy cost is paid at the point of generation, not at the point of use (since the end use of electricity is clean). The 10% environmental energy increase will increase the heat rate by 10%, to 11,000 Btu/kWh. At this

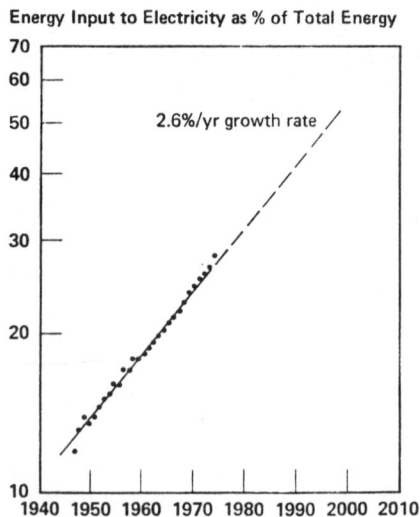

Fig. 14. Relationship between electricity and total energy in the United States.

heat rate 81.7 quads of fuel will be required for the generation of electricity in the year 2000. Note that the energy costs associated with environmental improvement increase the planning target fuel requirements (from 74.3 quads to 81.7 quads) and the heat rate (from 10,000 Btu/kWh to 11,000 kWh) but do not change the electricity production target.

The estimate of the electric fraction is in general agreement with the IEA estimate for the percentage of fuels used to produce electricity although the actual amounts differ due to the difference in total energy requirement estimates. The fractions calculated by the IEA Study are 50.8% for the high case (64 out of 125.9 quads) and 46.6% for the low case (47.3 out of 101.4 quads). The von Hippel-Williams analysis forecasts the electric fraction to be 39.8%.

The stated growth rates for electricity growth are 4.8% and 3.5% per year for the high and low cases respectively. These estimated growth rates coupled with the stated fuel requirements indicate the heat rate for the year 2000 to be between 10,300 Btu/kWh and 10,600 Btu/kWh for the IEA Study. The use of a heat rate in this range would drop the EPRI target by 3 to 5.2 quads. This difference is included in the extra environmental energy costs shown in Table 17. Lovins' scenario calls for a cutback in the use of electricity to about 15 quads by the year 2000. His total energy estimate for that time is roughly 95 quads.

Sources of Energy to Achieve the Planning Target

Research and development is underway for a number of new energy options. In the non-electric sector, two options are expected to make a measurable impact by the year 2000. The direct use of solar energy for heating and cooling is estimated to provide about 1% of the total energy in the year 2000. It is anticipated that this will replace oil and gas, because these are the primary heating fuels today. Some replacement of electricity by solar energy is possible, but this may be offset by increased use of solar-assisted heat pumps and the pumping requirements associated with solar heating. Gasification and liquefaction of coal will begin to displace oil and natural gas by 2000 and help reduce oil- and gas-import requirements.

New electric generation technologies that conceivably could be significant by year 2000 are but two, solar and geothermal. Except for dry-steam geothermal (a very scarce resource), neither of these is yet in the initial commercial stage; thus it is difficult to make a good estimate of their growth. Assuming very optimistic technical developments, 20,000–40,000 MWe of solar is a maximum possible range for the year 2000. Geothermal capacity of about 40,000 MWe by the year 2000 is projected by ERDA and EPRI.

New technologies are not expected to provide a substantial fraction of the year 2000 energy. To achieve a 150-quad planning target, the nation must continue to use the traditional fuels: oil, gas, and coal, and the developing fuel, uranium. The resource requirements under the assumptions of the planning target are shown in Table 18.

TABLE 18
Year 2000 Resource Distribution

	1973	EPRI planning target 2000 (10^{15} Btu/year)	IEA High growth 2000	IEA Low growth 2000	von Hippel-Williams moderate conservation case	Lovins 2000
Fuels for electricity						
Coal	8.7	37.7	20.0	7.1	17.5**	
Uranium	0.9	31.2	31.0	27.2	12.4	
Oil and gas	7.2	4.4	4.8	4.8	1.9	
Hydroelectric	2.9	4.4			3.7	
Geothermal, solar thermal electrical, wind	negl.	4.0	8.2	8.2		
Subtotal	19.7	81.7	64.0	47.3	35.5	15
Fuels for direct use						
Coal	4.7	5.8	11.2	10.6	5.2	
Liquid and gaseous fuels	50.3	59.1	48.7	41.5	46.8	
Solar heating and cooling and geothermal heat	negl.	2.0	2.0	2.0	1.7	
Subtotal	55.0	66.9	61.9	54.1	53.7	80
Total energy	74.7	148.6	125.9	101.4	89.2	95

*Lovins' breakdowns are approximately coal ~ 25q, oil and gas ~35q, and "soft" technologies ~ 35q.

**5.8 quads of this coal-fired electricity will be cogeneration.

Impacts on Oil and Gas Requirements

In the future, the electric fraction will represent to a great degree the extent to which coal and uranium have been substituted for oil and gas. These impacts are summarized in Table 19.

Clearly, the higher the electricity planning target, the lower the import requirements. Several things are worth noting about the other scenarios. The IEA year 2000 estimate projects 9.5 and 8.5 quads of shale and biomass for the high and low case respectively; the EPRI analysis does not plan for any fuel from these sources. The von Hippel-Williams study forecasts 6.2 quads from organic waste including methane from crop residues and manure, urban waste, and logging and pulp waste. The remarkable result is that of Lovins. He described a scenario in which oil and gas consumption increases from 57 quads in 1973 to 72 quads in 1985 and back down to 35 quads in the year 2000. With such a rapid decline in consumption described for post-1985, it is difficult to see how the investment will be made to develop a 72-quad/year handling capacity. Such rapid changes in capital intensive industries seem highly suspect — an oil company which believed this scenario

TABLE 19
Liquid and Gaseous Fuels (quads)

	Domestic supply	Imported fuel	Total
1973 actual	43	14	57
1985 IEA high	43.7	12.3	56
1985 IEA low	43.7	6.7	50.4
1985 Lovins	*	*	72.0
2000 EPRI (low electric fraction)	43	31.5	74.5
2000 EPRI (planning target)	43	20.5	63.5
2000 EPRI (high electric fraction)	43	7.1	50.1
2000 IEA high	47.3	6.2	53.5
2000 IEA low	46.3	0	46.3
2000 von Hippel	*	*	48.7
2000 Lovins	*	*	35.0

*Not given.

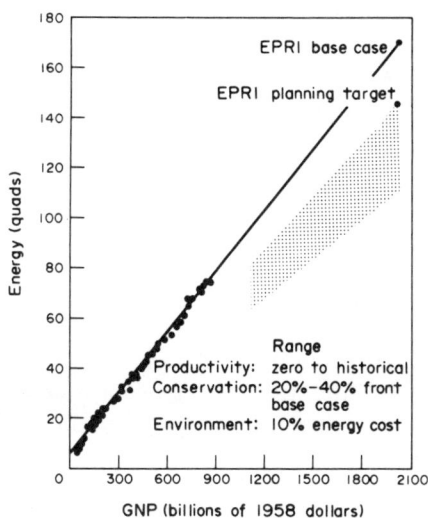

Fig. 15. Year 2000 energy and GNP, EPRI range.

is not likely to increase pipeline and refinery capacity in the early 1980s. In total this approach appears similar to calling for a massive expansion in buggy whip capacity in 1910.

The sensitivity of the planning target to all of the assumptions is illustrated in Fig. 15. The dotted area corresponds to the conservation range of 20–40%, with productivity varying from 1975 levels to that projected by a continuation of the historical growth and with a 10% environmental energy charge included. Because of the conservative nature of a planning target, the EPRI position is that it would be imprudent to plan for less than historical productivity growth or for more than 20% conservation. This corresponds to the upper corner: GNP is 2025 billion 1958$, energy is 150 quads.

Figure 16 includes the IEA and von Hippel forecasts and Lovins' scenario for comparison. (Lovins does not estimate GNP in the year 2000, hence the single line.) As mentioned previously, the difference between the IEA high case and EPRI planning target is due almost totally to different productivity growth estimates.

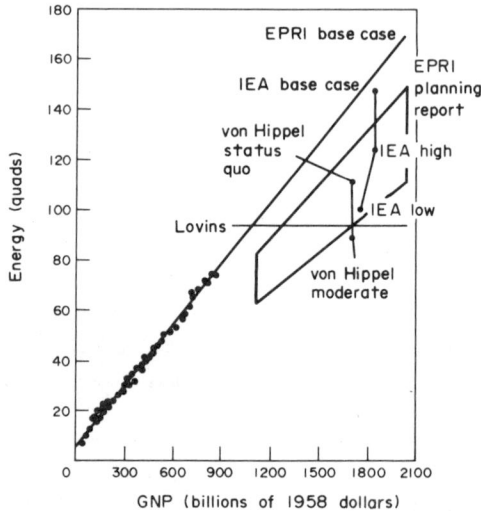

Fig. 16. Year 2000 energy and GNP, comparison of estimates.

Energy Analysis Assumptions — Comments

The different economic assumptions used in the IEA, von Hippel, and EPRI studies have been discussed; Lovins does not use a methodology based on the relationship between economic output and energy demand. Two other classes of assumptions are present in all four papers. Technological assumptions concerning the availability of future energy resources and energy options are quite important. The EPRI analysis assumes that little energy will be available by the year 2000 from advanced energy options — fusion, solar electric, etc — and with the exception of a projection of 9.5 quads of shale oil plus biomass fuels, the IEA study assumes likewise. The precise number for the EPRI study is that geothermal, solar of all types, and wind will provide 6 quads (out of 150). The IEA projects a similar number, plus the biomass and shale mentioned previously. The von Hippel-Williams estimate is similar; it projects little energy from new sources except for 6.2q from organic wastes. The technological question for these three analyses is: can the coal and nuclear levels projected for the future be achieved? The EPRI planning target assumes that coal production can almost triple in 25 years. The resource is certainly available; the limitations in this case would be expansion of the railroads, construction of mining equipment, availability of coal miners, and construction rate of coal-fired power plants from the 1973 capacity of 167 GWe to roughly 680 GWe. The Task Force on Energy of the National Academy of Engineering *(Energy Prospects: An Engineering Viewpoint, 1974),* after reviewing all of the problems associated with increasing coal production and delivery, reported: "Despite all of these problems, the Task Force felt that given sufficient incentives, it is within the capability of the coal industry to expand mine production by about 660 MTPY (million tons per year) in the next 11 years (1974–1985). . . ." At that time coal production was 600 MTPY, so a doubling of coal production by 1985 was deemed possible.

Assuming this is valid, and further assuming that this rate of expansion could be maintained until the year 2000, a six-fold increase in coal production would be possible. At this production level all of the electricity (at planning target levels) could come from coal. Significant disincentives exist for such rapid development; given the availability of other sources and a desire for a diverse array of production methods a three-fold increase seems more reasonable. Similarly, if the NAE nuclear capacity estimates of achievable construction rates are valid the nuclear planning target (32.1 quads and 500 GWe capacity) could be exceeded by a large margin. The NAE estimate (made in 1974) was that 325 GWe capacity by 1985 was possible. This would require about 30 GWe capacity per year coming on line in the late 1970s and early 1980s. With no increase in this rate 800 GWe by 2000 would be possible.

Lovins' primary technical assumption is that the developing technologies, wind, solar, biomass, will require less investment than the conventional nuclear and coal technologies. He further assumes that these technologies will be available when needed, with operational characteristics and sufficient reliability to provide for widespread public acceptance.

The capital cost estimates in both Lovins' papers have been criticized as biased towards the solar system he endorses. In a letter to the editor of *Foreign Affairs,* Hans Bethe argues that the cost estimates for solar heating are low by a factor of 4–5 (or greater if 100% solar heating capacity is assumed) and nuclear investment requirements overestimated by about a factor of 3. Further, the comparison between solar and electric heat should also include the electric heat pumps which give 2–3 times the heat value as resistance heating. The direct use of coal for home heating is a part of the Lovins' scenario. The required environmental standards would be met by the use of small fluidized-bed combustors. As proof of the maturity of the technology, he cites one instance in which a European firm has offered to build a large fluid bed combuster. The lack of operating experience in commercial operation coupled with the poor reliability and high maintenance experience with experimental fluidized beds makes this projection highly suspect.

Technical Factors

Although this has been discussed to some extent previously, it is useful to comment on some events which might occur that would effect severe changes in potential energy policies. The EPRI planning target, IEA, and von Hippel-Williams' forecasts rely almost entirely on proven technologies; for this reason unanticipated technical factors are unlikely to disturb future plans based upon these energy futures. The future uncertainty of technological capability profoundly affects the Lovins scenario. It should be remembered that any technology will become less attractive when it approaches commercialization because unforeseen difficulties are always encountered. Several years ago MHD (magnetohydrodynamics) was considered a promising method for using coal in a clean and efficient way; today the prospects have dimmed because the "engineering details" required to produce the required reliability have proven extremely difficult. So it may be with the small fluidized bed combustors in the Lovins scenario. The optimistic cost estimates for nuclear power in the 1950s and ocean thermal systems in the 1960s confirm that the early developers of a technology tend to become overly optimistic; a similar escalation of solar energy costs would severely impede its development. Failure to develop a cheap reliable source of storage might disqualify the use of wind power.

Increasing worldwide food demands could stop biomass programs, with the land, water, and fertilizer needed for these fuels going to much better use for food production. The rapid growth of these "soft" technologies in the Lovins scenario cannot tolerate any delays, let alone technical failures. Technical delays at present are balanced by increased oil imports, but in his scenario the oil-handling capabilities are already over-extended in the mid-1980s.

Economic Factors

As was indicated earlier, different assumptions concerning future productivity are responsible for most of the difference between the EPRI and IEA studies. It is useful to note that the 4% annual growth in real GNP anticipated by President Carter (and necessary to meet his economic objectives) is higher than that projected by any of these studies. A more precise estimate of this key variable is not available because the systems and discoveries that will cause productivity to increase are not yet known (if they were, they would be implemented immediately).

Capital availability is the other economic factor which could have far-reaching impacts on energy system expansion. Current estimates are that sufficient capital will be available. (In a GNP-energy linked system, the available capital grows with GNP as does the capital required for energy systems.) If sufficient capital is not found, energy policies based upon any of these scenarios will be in trouble. The rapid fluctuations in energy sources described by Lovins would lead to faster amortization of energy facilities. Typically, large energy-producing facilities are expected to last from 30 to 40 years, but this would not be the case in Lovins' scenario (oil and gas would rise to 72 quads in 1985, drop to 35 quads in 2000) and all nuclear capacity would be closed immediately. The costs associated with this transition seem not to have been included in his scenario.

Social Factors

A number of social factors will have significant impact on future consumption, conservation and environmental costs as discussed earlier. The other factors can be broken down into several areas:

Electrification. The EPRI planning target and IEA analysis project rapid expansion of electrical systems, von Hippel slightly less than these two and Lovins described a reduction from present levels and abandonment of the large central station generation system now in use. Certainly electricity was becoming the fuel of choice even before future delivery of oil and gas became questionable. The unique convenience and point of use cleanliness and efficiency contributed substantially to its rapid growth. It is the scale of the electrical system that is the source of Lovins' objection.

> "In an electrical world, your lifeline comes not from an understandable neighborhood technology run by people you know who are at your own social level, but rather from an alien, remote, and perhaps humiliatingly uncontrollable technology run by a faraway, bureaucratized, technical elite who have probably never heard of you."

Whether this bothers the public at large as much as it bothers Mr. Lovins is highly suspect, as large "remote" electrical systems have been in operation for 50 years or so without any apparent large-scale dissatisfaction.

A social factor not discussed by Lovins is the probable response to very small systems, perhaps diesel generators for every few houses. The loss of reliability with such a system as well as higher cost, noise, very irregular load patterns resulting in frequent capacity shortfalls, and required personal maintenance would seem to be strong impediments to this system. Those who live in apartments which share water heaters with neighbors can attest to some of the problems of "neighborhood" energy systems. The generation system described by Lovins which uses hydro, industrial cogeneration, small wind stations as well as local diesels certainly dilutes decision-making capability, but it also dilutes the responsibility which today provides for reliability of services.

Lifestyle. Lifestyle changes which could have an impact on energy demand are difficult to analyze. At one extreme is the notion that people will become and remain conservation conscious, that thermostats will stay at 65°F or lower, car pools will increase, and energy-intensive activities diminish. Counterbalancing this prospect is the projection that real income will increase dramatically, and that with these larger incomes will come increased demands for virtually all amenities. It is not possible to predict with any degree of confidence how people will spend their money, but it is certain that each dollar spent will result in the use of energy. The "energy intensity" of the average purchase will determine demand. It is probably not an over-statement to argue that the long-term trend represents increasing demand for comfort and convenience — this is certainly true for appliances: witness the growth of self-cleaning ovens, frost-free refrigerators, air conditioning, and increased residential lighting; for automobiles: until the oil embargo such features as automatic transmission and air conditioning were increasing in popularity; and in food, as indicated by the recent rise in popularity in prepared foods and fast-service restaurants. With incomes projected to roughly double, the higher energy costs may not offset the continuation of this trend. No analyst can claim any prescience in this area, but the rapid return to large automobiles following the oil embargo suggests that it would be imprudent to depend on a great deal of this type of conservation.

Social goals. It seems extremely likely that a strong national goal for the balance of the century will be the improvement of the conditions of the poor. The approach taken will produce substantial repercussions for energy demand. The manner through which such improvements will occur can vary between two approaches. One way would be by redistribution of income, and a reduction of the spread between the wealthy and the poor; the second would be to strive for rapid economic growth, with all incomes rising accordingly. This second approach comes closer to describing the historical trend. Should this trend change dramatically in favor of income redistribution with lower overall growth, energy growth rates would drop.

This distinction serves to define the philosophical difference between Lovins (and Schumacher) and more traditional analysis. For the purpose of an example, happiness (as it relates to material goods) can be equated to consumption divided by desire, with the ratio always less than one. The traditional approach is to increase consumption through productivity increase. The Lovins-Schumacher approach is to hope for a reduction in desire. Whether such a change in attitude can and will occur is an unanswerable question, but as future events develop, the appropriate path will be indicated.

11

Comments on "The National Energy Plan"

Part 1. STRATEGIES IN ENERGY PLANNING

"In developing public policy toward the energy crisis, all three possibilities — the most likely case, the optimistic case, and the pessimistic case — should be considered. It would be foolhardy to base public policy on the most optimistic possibility. Even if the future should prove to be brighter than now appears likely, steps taken to curb demand and increase use of abundant resources would still have been justified to meet the immediate need to reduce vulnerability. In formulating public policy toward energy, the prudent course is to act on the basis of the most likely assumptions about the future, and to bear in mind that the pessimistic set of assumptions is a real possibility."

> *The National Energy Plan*,
> Executive Office of the President,
> Energy Policy and Planning,
> 29 April 1977

The above statement must be regarded as the foundation on which the Administration's energy plan is based. The intent of this paper is to examine one single question: If we "act on the basis of the most likely assumptions" are we in fact following "the prudent course"?

To examine this question in the simplest way, only one factor is needed: the available capacity minus the capacity demanded at some point in the future. Because neither the capacity nor demand can be forecast exactly, this difference can be examined in a probabilistic sense.

Curve A of Fig. 1 represents the probability density of the difference between capacity and demand mentioned above. This distribution is represented in somewhat of an arbitrary manner, subject to the constraint that the most likely estimate of demand is equal to the most likely estimate of available capacity. Curve B represents the outcome if policies were adopted in which a surplus was the most likely outcome; this could occur if the policy were modified to provide increased supply capacity or increased conservation from curve A.

The central hypothesis of this analysis is that the social cost of a mismatch

---

* Presented at the Conference on National Energy Policy, Washington, DC, 17 May 1977.

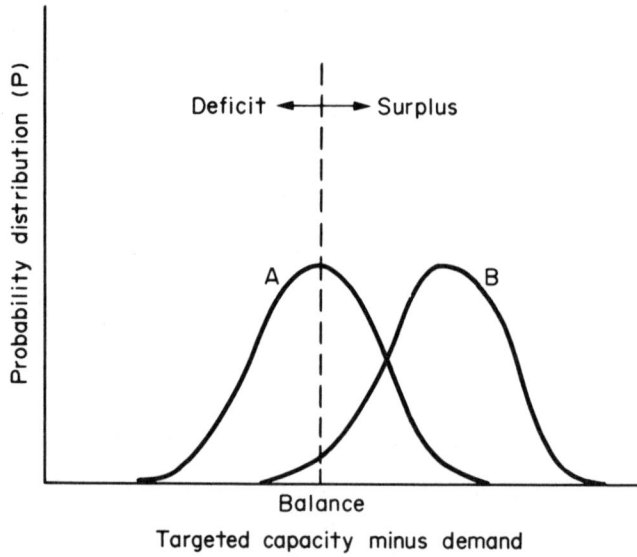

Fig. 1. Probability distribution of targeted capacity minus demand. Curve A: capacity equals demand, "most likely outcome" approach. Curve B: capacity surplus.

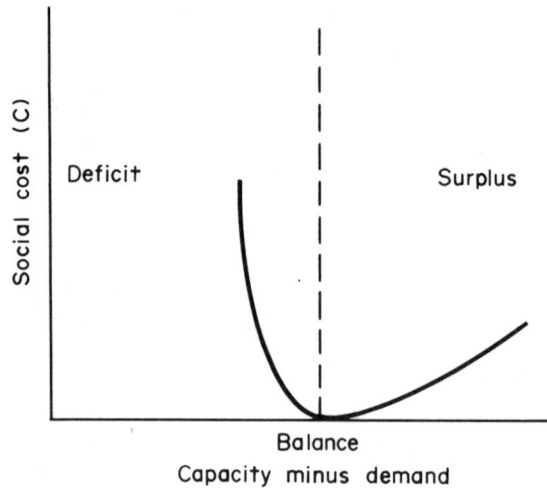

Fig. 2. Social costs of supply mismatch.

between energy capacity and capacity demanded is highly asymmetrical. Figure 2 represents the hypothesis that an energy capacity deficit is far more costly than a capacity surplus of equal size. This hypothesis is well supported by a February 1977 Stanford Research Institute Report, *Decision Analysis of California Electrical Capacity Expansion*, in which a number of curves similar to that of Fig. 2 are found. Figure 3, taken from this report, makes this point. In Fig. 3 the social cost is displayed as a function of the chosen capacity expansion rate. The social cost for

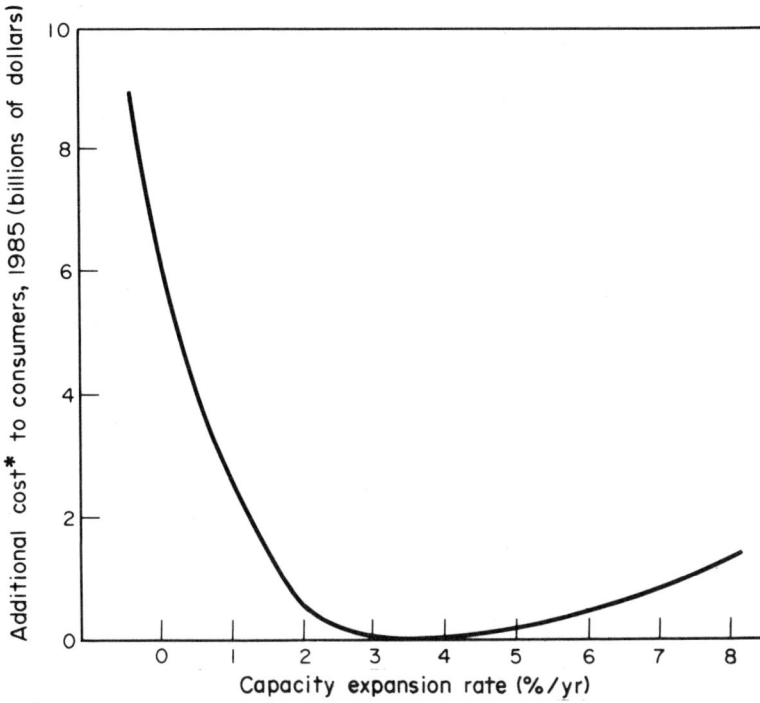

*Includes outage cost

Fig. 3. Expected value of relative consumer benefit (illustrative data).

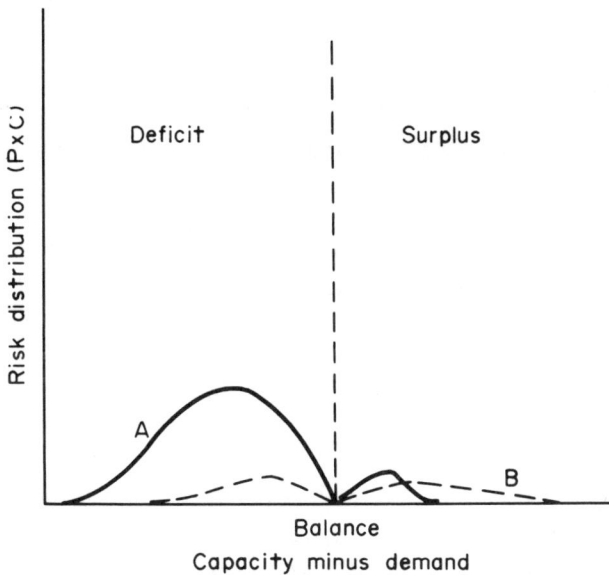

Fig. 4. Weighted social cost versus level of mismatch. Curve A: "most likely outcome" approach. Curve B: capacity surplus. *Note:* Expected cost equals areas under curves.

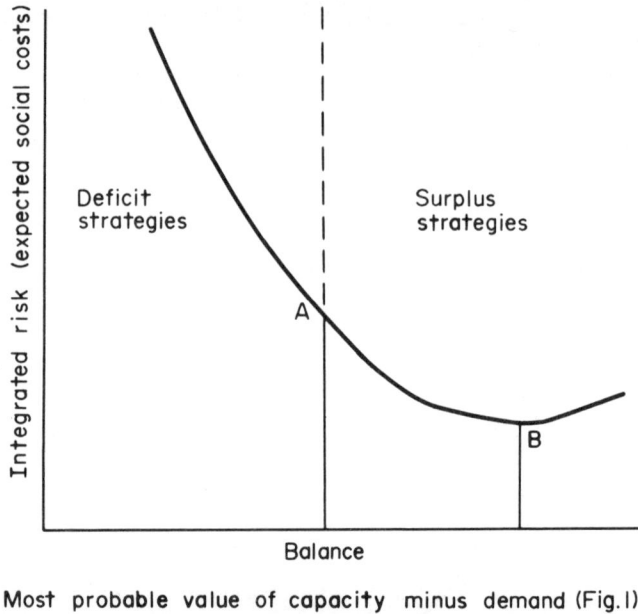

Most probable value of capacity minus demand (Fig. I)

Fig. 5. Expected cost versus strategy.

low growth rates is due to outages; for high growth rates the cost is due to surplus
capacity. The SRI report states:

> "*Insufficient capacity appears to impose costs on the consumer greater than those from
> excess capacity.* This is because outages are more likely when the system has
> insufficient capacity and because the losses from outages appear to be
> significant. Price-induced demand reductions due to overbuilding appear to
> be less of a social burden than possible outages due to underbuilding."

The weighted cost or risk distribution (social cost times probability) as a function
of the degree of mismatch is shown on Fig. 4. The integral of this curve (or the area
under it) represents the expected social cost due to a mismatch. It is clear that, for
the "most likely case" (curve A), the preponderance of the risk is due to a capacity
deficit.

Figure 5 represents this expected cost, or the integrated risk, as a function of the
most likely outcome chosen for capacity minus demand (Fig. 1). It is apparent
that the "most likely" supply and demand strategy does not produce the lowest
expected cost, and is therefore not the most prudent case.

Improvements in Energy Planning

Energy planning which reduces the likelihood of a shortfall was shown above to
be preferable to a strategy based upon a balance of most likely demand with
capacity. There is an additional benefit of increased domestic capacity, as indicated in Fig. 6. As stated in *The National Energy Plan*,

"To the extent that electricity from coal is substituted for oil and gas, the total amounts of energy used in the country will be somewhat larger due to the inherent inefficiency of electricity generation and distribution. But conserving scarce oil and natural gas is more important than saving coal."

It is quite clear that for a given level of energy supply, a supply mix which reduces imports is preferable for reasons relating to balance of trade and security of supply. Figure 6 represents this point. If a policy to increase domestic production is developed and oversupply results, a likely outcome is a reduction of energy imports. *To the degree that substitution of domestic sources of energy for imports is permitted, there may be virtually no social cost associated with some amount of overcapacity.*

The shift in Fig. 1 from curve A to curve B might be characterized as a "build-like-mad" approach, for this shift represents a significantly increased energy capacity relative to some fixed level of demand. While it was shown that curve B produces lower expected costs than the "most-likely" approach, there are further refinements that can be made to reduce cost even further. The first of these is to improve the information used as a basis for planning.

The Impact of Uncertainty

If future capacity and demand were known with certainty, the "most likely outcome" approach would be correct, and social cost could be held to a minimum. As uncertainty increases, so does the desirable amount of surplus, and so does the expected value of the loss.

This effect is perhaps best explained by an analogy to a checking account. If an individual knows what his expenses will be with precision for the next month, he could and would put exactly that amount of money into his checking account. The remainder of his money would stay in savings to earn interest. But because

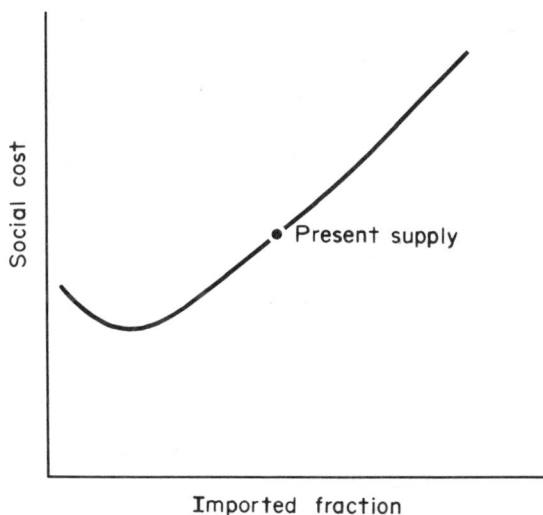

Fig. 6. Social cost as a function of imports.

expenses are rarely certain, he puts more money into his account than he actually expects to spend. A balance is sought between the high probability or certainty that he will lose some added interest with the low probability that he will suffer the high social cost of the embarrassment of being overdrawn. Similarly, it is in the best interests of the nation to pay for a small amount of excess energy capacity to reduce the chances of a costly energy shortage.

This points up the extremely high value of information. Reduction in uncertainty, both of capacity available and demanded, provides corresponding reductions in the size of the desirable cushion.

Improvements in the Flexibility of Energy Policy

Response flexibility relates to the dynamic nature of the problem. If, in 1977, plans are adopted which minimize the expected loss relative to the presently available information, these plans will probably not be the best choice in 1980. The passage of time brings new information, and a policy should be flexible enough to utilize new information while recognizing that extreme fluctuations in policy will produce uncertainties which carry their own social costs. One policy of the Administration's which increases response flexibility is the concept of a strategic petroleum reserve. This buys time in the event of a short-term undersupply, yet is significantly cheaper than developing an equivalent production capacity. A 4 billion barrel storage supply costs about $60 billion. In an emergency, this reserve could supply 11 million barrels per day for a year. If $60 billion were spent on electric power plants, the 60-GWe capacity is equivalent to roughly 1½ million barrels a day, but is available for decades rather than for 1 year. The storage reserve is an emergency insurance only.

What is needed to completely utilize the concept of strategic storage is a list of complementary contingency plans which would be implemented when it appears that an energy or capacity deficit is approaching. On the demand side, an emergency fuel allocation plan should be established which could minimize the costs of supply disruptions. For the supply side, policies that would permit rapid increases in capacity are needed.

The value of reduced plant construction time is now seen to provide benefits in two separate ways. In addition to the direct-cost reductions produced when lead time is reduced, the response time of the whole system is improved so that the mismatch probabilities (Fig. 1) are reduced.

For the longer term, research and development of alternative energy options provides this to some extent. In this light, the decision to slow the breeder reactor program appears highly questionable. In the case of electrical capacity, the probability of a deficit is due to uncertainties in demand growth, the ability to expand coal production and delivery, and the supply of uranium. If future events indicate that the present "best guesses" in these areas are seriously in error, then the present judgment about the desirability of the breeder will certainly be questioned. Development of the breeder now to the point at which it is a future option greatly increases the flexibility of future policy. While this technical development is proceeding, additional information concerning uranium

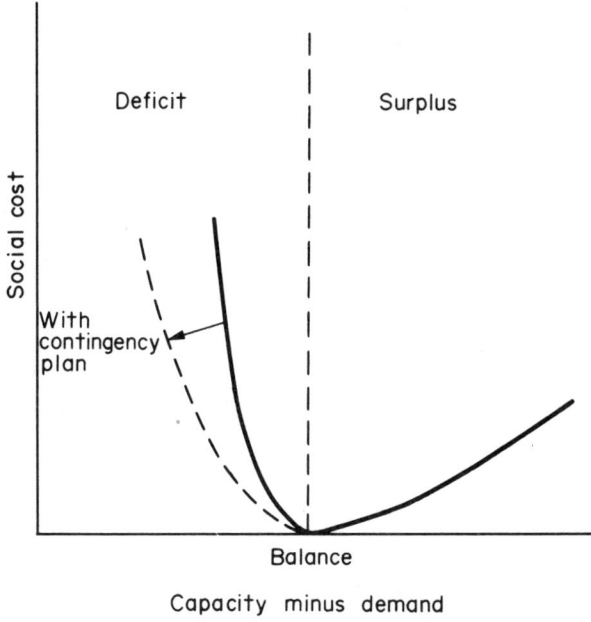

Fig. 7. Impact of improved shortfall response.

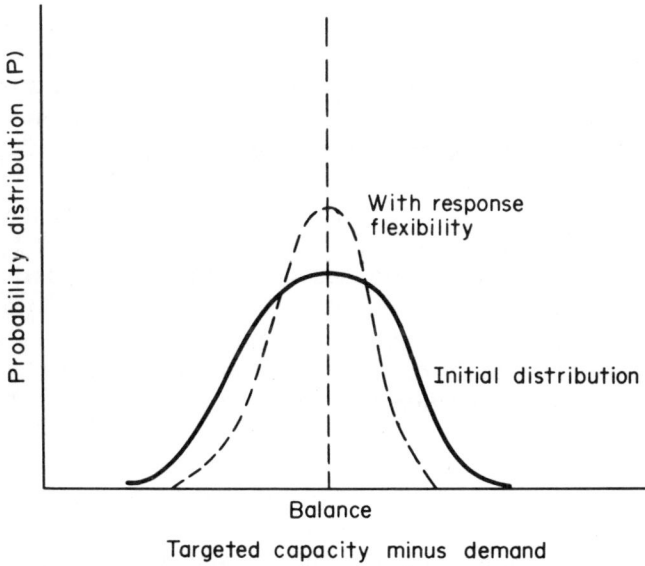

Fig. 8. Impact of capacity response flexibility.

resources, technical characteristics of breeder systems, particularly as they relate to proliferation, and the policies of other nations can be developed. At some point in the future this additional information, coupled with updated forecasts of energy supply and demand, will provide the basis for decisions concerning commercial breeder systems.

These policies can be regarded as providing strategic insurance. The effect of a demand contingency plan on social cost is indicated in Fig. 7. The supply flexibility results in the shift shown in Fig. 8.

Conclusions

The proposed energy plan, based on the "most-likely assumption", results in a high societal risk. The most prudent policy (a stated goal of the plan) would be one in which a capacity shortage is much less likely. Several features of a policy which minimizes risk are: (1) the plan will provide for somewhat more capacity than current best estimates suggest will be needed, (2) actions which improve response flexibility will be adopted. These latter actions include the use of storage, reductions in the lead times for energy source development, research into alternative energy options, and well-thought-out plans to reduce demand and allocate available fuel given a shortfall.

Finally, a desirable policy will recognize the value of trend information, and will be responsive as new information becomes available.

PART 2. NUCLEAR POWER AND PROLIFERATION

"Proliferation" — the current shorthand buzzword which describes the potential international spread of the production capability for nuclear weapons — has been a matter of national concern in the US since the closing days of World War II. But this recently popularized concern now appears to be causing a reversal of a quarter-century of US policy regarding the best means of preventing proliferation. Stimulated by the last presidential campaign, the US has been moving toward prohibiting, or severely restricting, domestic use of the civilian fuel cycle including plutonium reprocessing and postponing consequently the US breeder reactor option. This is being advocated on the ground that if the US foregoes civilian reprocessing, use of plutonium, and delays the breeder, other countries — energy-hungry though they may be — will voluntarily deprive themselves of the full benefit of nuclear energy to follow our "moral leadership".

The Administration's recent policy announcement on nuclear power has the effect of promoting LWRs and delaying both plutonium recycle and the breeder. The basic premise of this policy is that a combination of known coal reserves and uranium resources yet to be found permit our future energy needs to be met through the turn of the century without recycle or the breeder. The LWR fuel would pass through the reactor only once, and then be stored indefinitely. Because the uranium resource future is uncertain, many of us believe the insurance aspect of the breeder option, which relates directly to Part 1 of this paper, should be fully developed now and subsequently used as required. All other advanced fuel cycles which give high utilization of uranium or thorium also require reprocessing. The Administration's counter-argument is that closing the plutonium fuel cycle, as required for a fully developed breeder system, would place the US in the position now of endorsing plutonium recycle, and thus encourage the development in other nations of a possible channel for supplying weapons material. It is also the Administration's contention that domestic pursuit of the breeder and recycle for US energy supply and simultaneous discouragement of other nations would create an unacceptable double standard — although such already exists in the weapons field. Many of us have a deep concern that the Administration's position on plutonium recycle and the breeder will be internationally counter-productive and actually stimulate proliferation, and may domestically damage our economy as well.

How real is the danger that reprocessing of civilian fuel would be used by nonweapons countries to obtain plutonium for weapons? A simple analysis of well-known facts shows that there are today no fewer than eight different ways available (Fig. 9) to produce weapons material — to *produce*, not steal. Among them, the route of commercial nuclear power using slightly enriched-uranium fuel ranks eighth and low in desirability for a country that has made a political decision to establish a nuclear weapons capability. All the weapons that exist today were made by other means. The one made by India was made from a research reactor; this weapon was the one which startled the world and gave rise to some of these concerns. But the research reactor is so simple that Argentina is now selling one to Peru.

Commercial nuclear power is the most expensive route to weapons development (five to ten times more costly), requires the highest level of support technology and the broadest base of support industry, and takes the longest to install and to yield material (3 to 5 years longer).

By the 1980s, moreover, the world may have available several new, additional ways (Fig. 10) of producing weapons-grade fissionable material, adding even more routes to weapons capability. Several of these are likely to be easier, cheaper, and faster for a nation bent on bootstrapping itself to acquire weapons capability than is the route of nuclear fission power with reprocessing and recycling of fuel.

The philosophical basis inherent in the Administration's nuclear policy is contained in the quote from Mr. Nye in Fig. 11. It is specifically the last point on

	Required		
	Cost	*Technology*	*Industry*
Research reactor	Small	Small	Small
Production reactor	Medium	Medium	Medium
Power reactor	Large	Large	Large
Diffusion cascade	Large	Large	Large
Centrifuge cascade	Medium	Medium	Medium
Aerodynamic jet cascade	Large	Medium	Large
Electromagnetic separation	Medium	Large	Medium
Accelerator	Medium	Medium	Medium

Fig. 9. Eight-fold ways available for weapon material production.

U-235 SEPARATION
- Laser
- Chemical exchange
- Jet membrane

Pu or U-233 PRODUCTION
- Plasma fusion-fission
- Inertial fusion-fission
     Laser implosion
     Electron beam implosion
- Accelerator (I.N.G.)

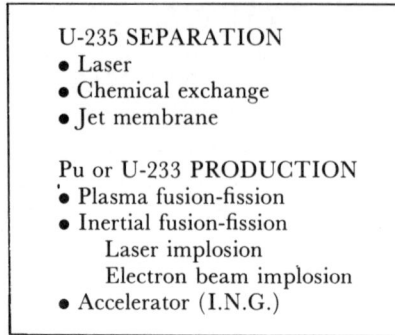

Fig. 10. Potential added routes to nuclear weapons materials (in the 1980s).

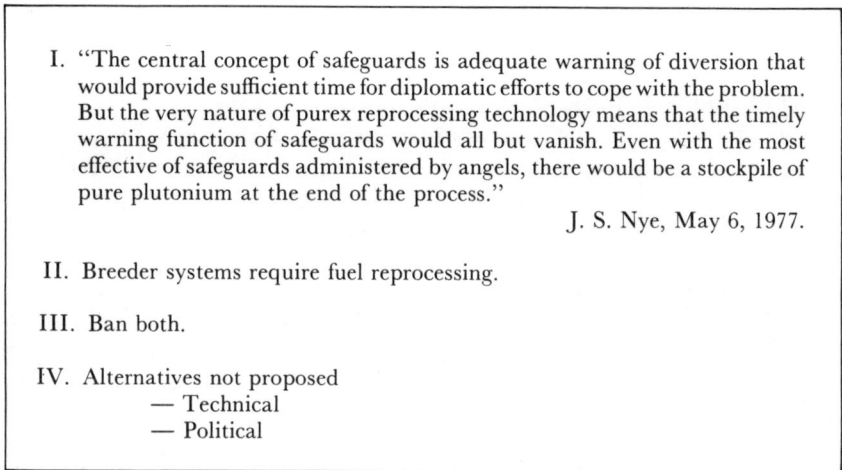

I. "The central concept of safeguards is adequate warning of diversion that would provide sufficient time for diplomatic efforts to cope with the problem. But the very nature of purex reprocessing technology means that the timely warning function of safeguards would all but vanish. Even with the most effective of safeguards administered by angels, there would be a stockpile of pure plutonium at the end of the process."

J. S. Nye, May 6, 1977.

II. Breeder systems require fuel reprocessing.

III. Ban both.

IV. Alternatives not proposed
        — Technical
        — Political

Fig. 11. Basis for present policy.

this figure, the considerations of technical and political alternatives that can lead to a policy that is sound on both proliferation and domestic energy supply issues, that is to be addressed here.

The technical alternatives for a breeder reactor lead to the same kind of machinery (Fig. 12) whether it is fueled by U^{233} from thorium, U^{235}, or plutonium.

The differences between these cycles is that a breeder on a U^{238}-plutonium cycle produces fuel roughly twice as fast as does the thorium-U^{233} cycle. This difference in fuel production rates gives the plutonium cycle a substantial economic advantage, as it can feed light-water reactors or provide fuel for the initial cores of new reactors at a much faster rate than can the thorium cycle machine.

In terms of proliferation, the thorium cycle buys nothing, as U^{233} is a material from which weapons can be made. These two factors indicate a strong technical advantage for the plutonium cycle breeder. The focus of proliferation concerns is

POWER PLANT

Turbine

93%

E
N
R
I
C
H

20%

U
Ore

4%

REPROCESSING

Th Base

U Base

Core

LWR

Pu

Spent Fuel

Pu

U^{233}

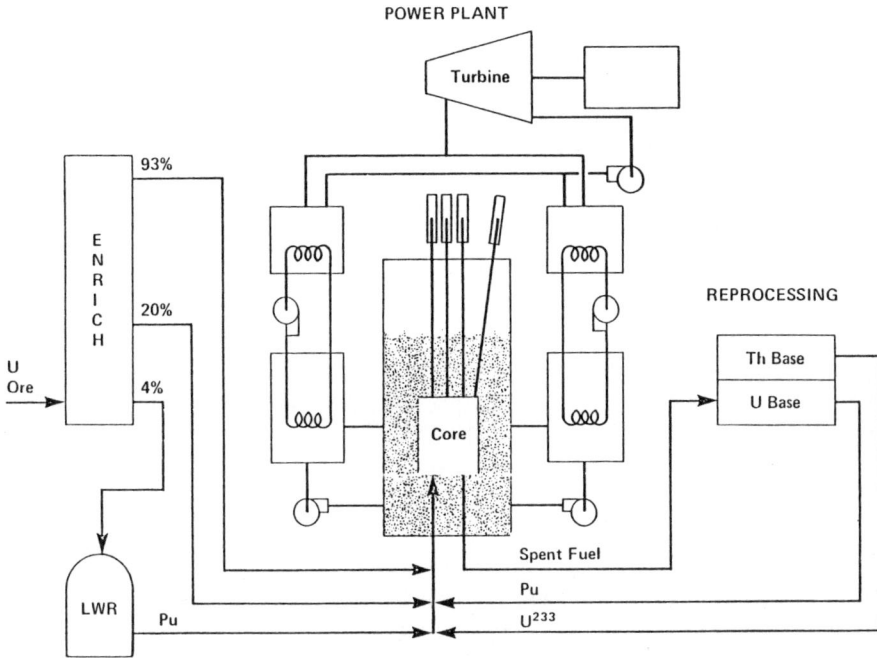

Fig. 12. Breeder fuel cycles.

upon the back end of the fuel cycle — the reprocessing and storage of reactor fuel. What has been forgotten by the people not in the nuclear business for the last 25 years is that the requirements for the reprocessing cycle for the fast breeder are quite different than what they are for the light-water reactor.

	Output	Quality	Appropriate process
Military weapons	Pu	Highest purity (weapons efficiency)	Purex
LWR	Pu	High purity (glove-box handling)	
	U	Highest purity (reenrichment)	Purex
	Fission products	Bulk removal (neutron economy)	
FBR	Mixture (Pu, U, F.P.)	Low purity (remote fabrication) Partial decontamination (fuel reconstitution only) High radiation level and short cooling times	Pyro-processing

Fig. 13. Technical reprocessing objectives.

In the light-water reactor the plutonium, uranium, and fission products must be separated out with high purity because glove boxes are used for new fuel fabrication. Additionally, the slightly depleted uranium sent back to the enrichment plant must be of highest purity. The requirements for purity are even more stringent for collection of weapons grade material. Military requirements are for the purest possible process and that was the purex process; it gives plutonium 99½% pure on the output. The purex process has also been the best method of meeting the reprocessing requirements for light-water reactors.

The requirements for the fast-breeder fuel reprocessing are a great deal less stringent than for these other two systems. In the breeder, two or more new atoms are formed as each fuel atom undergoes fission so the volume of the material actually expands. In reprocessing, the bulk of the fission products are removed just to account for this physical expansion; the plutonium can be of extremely low purity, because remote handling is required (Fig. 13).

For this application the appropriate method is pyro-processing, described in Fig. 14.

The point is that there are technological means of continuing the nuclear power development, the breeder and the chemical reprocessing, that remove pure plutonium from the cycle and make it very difficult to get pure plutonium for weapons purposes.

Probably one of the most hazardous things is the very recommendation the Administration is making. Light-water reactor fuel that has been stored for 5 or 10 years can be processed by any chemical engineering institution including universities — all they need is a backyard swimming pool to cut the material up.

The pyro-processing technology, which has already been demonstrated, offers greater resistance to proliferation than does the method recommended by the Administration.

The principal point is that there is no justification of national policy for the Administration to hold up the development of one of the major resources, the fast breeder with plutonium recycle. What the Administration should have done was to come to the technical fraternity of the US to find technologies to use this resource which removes the proliferation opportunities which the present cycle provides. That was not done.

(Non-aqueous molten salts and metals to separate constituents)

I. Early studies (1940s) revealed inadequacy for weapons grade Pu.

II. Unique advantages for FBR fuel.

III. Background research already exists.

IV. Complete cycle demonstrated on EBR II (1963–1968).

V. FBR with pyro-processing more "proliferation resistant" than proposed permanent storage of LWR once-through fuel.

Fig. 14. Pyro-processing.